THE RISE AND RISE OF INDICATORS

This book makes indicators more accessible, in terms of what they are, who created them and how they are used. It examines the subjectivity and human frailty behind these quintessentially 'hard' and technical measures of the world.

To achieve this goal, *The Rise and Rise of Indicators* presents the world in terms of a selected set of indicators. The emphasis is upon the origins of the indicators and the motivation behind their creation and evolution. The ideas and assumptions behind the indicators are made transparent to demonstrate how changes to them can dramatically alter the ranking of countries that emerge. They are, after all, human constructs and thus embody human biases. The book concludes by examining the future of indicators and the author sets out some possible trajectories, including the growing emphasis on indicators as important tools in the Sustainable Development Goals that have been set for the world up until 2030.

This is a valuable resource for undergraduate and postgraduate students in the areas of economics, sociology, geography, environmental studies, development studies, area studies, business studies, politics and international relations.

Stephen Morse is Chair in Systems Analysis for Sustainability at the University of Surrey, UK.

THE RISE AND RISE OF INDICATORS

Their History and Geography

Stephen Morse

LONDON AND NEW YORK

First published 2019
by Routledge
2 Park Square, Milton Park, Abingdon, Oxon OX14 4RN

and by Routledge
52 Vanderbilt Avenue, New York, NY 10017

Routledge is an imprint of the Taylor & Francis Group, an informa business

British Library Cataloguing-in-Publication Data
A catalogue record for this book is available from the British Library

Library of Congress Cataloging-in-Publication Data
Names: Morse, Stephen, 1957- author.
Title: The rise and rise of indicators: their history and geography /
Stephen Morse.
Description: Abingdon, Oxon; New York, NY: Routledge, 2019. |
Includes bibliographical references and index. |
Identifiers: LCCN 2019002662 (print) | LCCN 2019012970 (ebook) |
ISBN 9781315226675 (Master) | ISBN 9780415786805 (hbk) |
ISBN 9780415786812 (pbk) | ISBN 9781315226675 (ebk)
Subjects: LCSH: Social indicators. | Environmental indicators. |
Political indicators. | Quality of life–Evaluation. | Sustainable
development–Statistical methods. | Sustainable Development Goals.
Classification: LCC HN25 (ebook) | LCC HN25 .M65 2019 (print) |
DDC 306–dc23
LC record available at https://lccn.loc.gov/2019002662

ISBN: 978-0-415-78680-5 (hbk)
ISBN: 978-0-415-78681-2 (pbk)
ISBN: 978-1-315-22667-5 (ebk)

Typeset in Bembo
by Deanta Global Publishing Services, Chennai, India

 Printed in the United Kingdom
by Henry Ling Limited

CONTENTS

FIGURES

TABLES

ABBREVIATIONS

CA	content analysis
CI	confidence interval (a term used in statistics)
CPI	Consumer Price Index, a measure of inflation (Chapter 2)
CPI	Corruption Perception Index
DAC	Development Assistance Committee
DMSP	Defense Meteorological Satellite Program
EF	Ecological Footprint
EPI	Environmental Performance Index
EQV	Equivalence Factor (used in the calculation of the Ecological Footprint)
ESI	Environmental Sustainability Index
FAO	Food and Agriculture Organisation (United Nations agency)
GB	Great Britain
GDP	Gross Domestic Product
GFN	Global Footprint Network
gha	global hectares (units used by the ecological footprint)
GNI	Gross National Income
GNP	Gross National Product
GPI	Genuine Progress Index
HDI	Human Development Index
HDR	Human Development Report (produced by the UNDP)
HI	Happiness Index
HIPC	heavily indebted poor country
HPI	Happy Planet Index
IHDI	inequality-adjusted HDI
IMD	Index of Multiple Deprivation
IMF	International Monetary Fund

LN	natural logarithm (logarithm to the base e, where 'e' is Euler's number)
MDG	Millennium Development Goals
ND-GAIN	Notre Dame Global Adaptation Index
NEF	New Economics Foundation
NLDI	Night Light Development Index
NOAA	National Oceanic and Atmospheric Administration
ODA	Official Development Assistance
OECD	Organisation for Economic Co-operation and Development
PPP	purchasing power parity
r	correlation coefficient
RPI	Retail Price Index (a measure of inflation)
SD	Sustainable Development
sd	standard deviation (a statistical term)
SDG	Sustainable Development Goal
SDGI	Sustainable Development Goal Index
TI	Transparency International (the organisation that produces the Corruption Perception Index)
TID	Townsend Index of Deprivation
UK	United Kingdom
UN	United Nations
UNDP	United Nations Development Programme
US	United States (of America)
WCED	World Commission on the Environment and Development
WDI	World Development Indicators (report produced by the World Bank)
WDR	World Development Report (produced by the World Bank)
WEF	World Economic Forum
WWF	World Wide Fund for Nature

FOREWORD

Triffids as indicators

Indicators have been around for a very long time. Indeed, and this may be something of a surprise to the reader, I will make the case in this book that they have been around for as long as humanity; but much depends how we define them. Here are two definitions of 'indicator' taken from dictionaries that provide two quite different perspectives:

> A thing that indicates the state or level of something.
>
> *(Oxford Dictionary)*[1]

> An indicator is a measurement or value which gives you an idea of what something is like.
>
> *(Collins Dictionary)*[2]

We have come to think of indicators as numbers, as set out in the *Collins Dictionary* definition above and their use of the terms 'measurement' and 'value', but the *Oxford Dictionary* definition is actually a much broader one that seemingly goes well beyond the use of numbers. Indeed, we use our senses every day to "indicate the state or level of something". We may look out of the window in the morning to get an idea of what the weather is like, and use visual clues such as the clothing that other people may be wearing to give us an indication about what we should wear. When driving a vehicle, we use clues such as the behaviour of other drivers around us, such as whether they are speeding or driving erratically, and attune our behaviour to what they are doing. Indeed, in the UK we use the term 'indicator' for the orange lights on the side of the car that we use to tell other drivers whether we intend to turn the vehicle right or left. The dashboard of the vehicle also has many indicators that tell us about its state, such as whether the engine is overheating or the level of fuel. But it is not just about visual clues. Noises can also be indicators we use every day. The growling of a dog or the purring of a cat tell us something about the animal's state, the alarms of emergency vehicles on the road tell us that something has happened that they are responding to. Also, of course, the smell of smoke is an indicator that there is fire nearby and this may be a threat to safety. These observations and many like them are things we do every day without even thinking about it; they are instinctive as well as learned – from our families, friends and, frankly, experience (good and bad). When put like this, indicators are part of us.

While the examples I have given in the previous paragraph all sound obvious, and indeed they are, there are also many more subtle indicators that people can use for their very survival. For example, farmers and hunters, whose livelihoods are highly dependent on the environment, will use a multitude of clues to help them make decisions about what crops to grow and how to best grow them, as well as where to hunt and how to identify and track prey. I worked for many years with farmers in West Africa, mostly in Nigeria, and they often used clues such as the presence of particular plants growing wild to tell them something about the nature of the soil and what crop they should grow. Over time I began to learn these indicators from them. An example is a plant with the name *Chromolaena odorata* (see Figure 0.1), a shrub native to the southern US, Mexico and the Caribbean, which has become widespread throughout the tropics, through accidental means but also, ironically, via planned release to control other weeds. Like so many invasive plants, once introduced into new habitats and free of its natural enemies that keep it in check in the place where it is native, it can multiply rapidly and become a weed. *Chromolaena odorata* is not a particularly attractive plant, at least in my opinion, but it has one of the most intriguing lists of English common names that reflect this invasive and unwanted ability.

(a)   (b)

FIGURE 0.1   *Chromolaena odorata* leaves (a) and flower (b). (a) Name of creator: Ashasathees; Title: *Chromolaena odorata* by Ashasathees. (b) Name of creator: Jeevan Jose; Title: *Chromolaena odorata*. Both pictures are licensed under the Creative Commons Attribution-Share Alike 4.0 International license

Source: Wikimedia Commons.

Probably the most prevalent common name is Siam weed, but it is also known as Christmas bush, devil weed, kingweed, serpentine weed, paraffin bush, paraffin weed, communist green and, believe it or not, triffid plant in some parts of the world. There are not many plants given a name ('triffid') from a work of science fiction; triffid is the name given to man-eating plants from the post-apocalyptic novel (and subsequent movies and TV series) called *The day of the Triffids*, written by John Wyndham. The triffids "grow … know … walk … talk … stalk and KILL!", as shown in the movie publicity poster from 1962 (Figure 0.2). *Chromolaena odorata* does not look like the triffids of the movie and does not walk, stalk and kill – thankfully – but the name reflects the invasive nature of the plant.

FIGURE 0.2 Movie poster for the *Day of the Triffids* (1962)

Source: Wikimedia Commons; Author: Joseph Smith; Title: Day of the triffids poster; Link: https://commons.wikimedia.org/wiki/File:Day_of_the_triffids_poster.jpg.

Some of the other names, as well as 'odorata' in the botanical name, reflect the pungent smell you get once the leaves are crushed.

Local names for *Chromolaena odorata* often reflect the time when the plant first came to their area. In the Igalaland part of central Nigeria, for example, famers refer to the plant as Il'ame and Ilijabiti. Il'ame means 'he has come' and refers to the accession of the King of the Igala people on 2 November 1956, while Ilijabeti is a phonetic name for Queen Elizabeth II, who was crowned on 2 June 1953. People often associate plant names with events that happen around the time of the plant's first arrival, and *Chromolaena odorata* arrived in West Africa in the early 1950s.

While *Chromolaena odorata* is regarded as a damaging weed in the tropics, hence I suspect some of the colourful and unpleasant names listed above, it does need good soil to grow and indeed can improve soil fertility and reduce some soil-born pests. Thus, ironically, farmers in West Africa will often see the presence of this plant as indicative of a good place to cultivate even if they do not want it to be around while their crops are growing. Its presence 'indicates' to them that the soil has good fertility and fewer pests. They do not count the number of plants or assess presence in numerical terms, such as the extent of soil coverage or extent of foliage, but they will visually assess its presence and coverage. They are not measuring anything but using visual clues to tell them about "what something is like".

Even triffids can be indicators!

But I do have to own up and say that this book is not about triffids. My focus is mostly on indicators, as defined in the *Collins Dictionary*. Thus, I have to admit that I am more at the end of the spectrum where indicators are defined as numbers and in this book I cover a lot of them. My aim is to introduce these indicators for people who may not have much experience of them, but although there will be a lot of numbers I have intentionally tried to steer away from providing too much maths. Instead my focus is more on the history of the indicators, where they came from and why they were created, together with what they tell us of the world today. Having said that, a key lesson with all these more numerical indicators is that an awful lot depends on the assumptions that rest behind them. Changing those assumptions can make a very big difference to the message that is being conveyed. Also, while the book leans towards the more numerical interpretation of indicators, I will also come back to the more instinctive and perhaps human perspective in Chapter 10 – although triffids will not be part of that.

It is common to use the terms indicator and index interchangeably, and within the *Collins Dictionary* definition either term can be used. But, for geeks like me, these two terms are different. An index is typically defined as a single number that comprises a number of indicators. In effect, the separate indicators are mathematically combined, perhaps by adding them together and taking the average, to produce a single value (the index). Many of the examples covered in this book are technically indices rather than indicators, but I do admit that the distinction can be blurred and, for the messages I am conveying for the audience I have in mind, it probably does not matter that much whether the terms used are indicators or indices. But even so I have followed this convention through the book and that is why you will see indices.

This book covers a number of the more widespread and, in some cases, novel indices, but with a special focus on why they were developed, what was their purpose, how they have been promoted and what they tell us about the world. I have avoided a more technical discussion about the mathematics behind each of the indicators, although in some cases a degree of this technicality is necessary in order to illustrate the key assumptions that were made by the index creators and what could result from that. This matters a lot as changing the assumptions can generate very different messages. But I have sought to keep such 'technical-speak' to a minimum, although references and further reading have been included at the end of every chapter, so the interested reader can follow up on the details if they wish. I have also opted to focus on how the indices are related to each other rather than treat each of them entirely in isolation. After all, while each index is created and promoted by its own group of owners, those who consume and act upon the messages that are being conveyed by the indices may well be exposed to many such indices, as well as a host of other influences, to help inform their decision-making. Therefore, we must think in terms of an ecosystem of indices, such that consumers can relate to them or combine them in order to help them to understand the complexities that they are faced with. For this discussion it is necessary to introduce some statistical concepts, notably correlation and regression. I acknowledge that not all the readers of this book will have any particular knowledge of statistics, so I have included some background to them without going into the detailed mathematics.

I could have opted for many indices, and experience tells me that I do have to start the book with a brief justification as to why I have opted for the ones I have. I have published a number of books on indicators and indices over the years and almost always receive some critical comments from colleagues and reviewers about why certain indices have been left out in favour of others. There can be much passion here, which in my early days of writing about indicators and indices did surprise me. I often wondered, and still do, at the excited response about what are, at heart, quite technical and mathematical devices. It provides a salient reminder that while indices may appear to be technical and impersonal they are trying to capture 'something' (poverty, livelihood, health, education, crime, happiness, fairness) that matters a great deal to most, if not all, of us. Even within these numbers there is human soul. Hence, I must admit straight away that my choice of indices for this book is partly a personal one, some of my favourites are in here, but one that has also been driven by a desire to cover as many issues in the field as possible, including some potential future developments. I certainly do not wish to imply in any way that the ones I have selected are the best indices in any sense of being the most objective, scientific, popular or effective. Indeed, these latter terms will themselves be teased apart throughout the book.

This book has been written for those who may not have much experience of indices, but who wish to know more about how these tools have come about and what they tell us about the world. Thus, students from a variety of disciplines, from the social and economic to the physical and natural sciences, will find it of use, especially if they are having to think about engagement with others outside their field of expertise. Indeed, these days those who fund research in many countries often require that the work must have a positive impact within wider society. To achieve impact often requires an engagement with business people, non-governmental organisations, civil servants, politicians or perhaps the wider public; and inevitably there is the challenge of communicating complex ideas to people who are not likely to be specialists in the field. Indicators and indices are often part of this communication, with the intention that consumers of the indicators will use them to help bring about change and impact. You may not have started out thinking about this need to engage with non-experts, but you could well find yourself in that position, whether you like it or not. I hope that this book helps you with that thinking.

Notes

1 The *Oxford Dictionary* can be accessed via: https://en.oxforddictionaries.com/.
2 The *Collins Dictionary* can be accessed via: https://www.collinsdictionary.com/.

ACKNOWLEDGEMENTS

I would like to thank all those who supported and encouraged me in the writing of this book, but especially my family – Maura, Llewellyn and Rhianna. I would also like to thank my Head of Department at the University of Surrey, Professor Richard Murphy, for allowing me the space to work on this book among the many demands on academic time.

1

THE WORLD IN NUMBERS

Indicators are all around

"Love is all around" sang the Troggs in their hit of the late 1960s (followed by the perhaps better-known Wet Wet Wet version in 1994, which was the theme song for the highly popular movie *Four Weddings and a Funeral*), but the same is also true of indicators, even if they do not make, sadly, for such a catchy song title or for mega-sales of hit singles. Indicators are all around. It might surprise you to know that you – and indeed everyone else – use them every day of your life, although I am pretty confident that you don't say to yourself "ahh ... another indicator – my tenth of the day". So just what are these ubiquitous 'things' that are all around us like love? Well, at its most fundamental level, an indicator is nothing more than a clue – a signal that something is happening, or perhaps a suggestion that something will happen. Therefore, as I look out of my window, I can unfortunately see the sky getting darker with clouds and this is an indicator to me that it may soon start to rain. The fact that it is cold, despite being our so-called spring here in the UK, is indicated by the proportion of people I can see wearing coats outside. The proportion is not 100% , as some hardy souls are even wearing t-shirts and shorts, but the majority view over appropriate clothing gives me a clue that I will need to wear a coat when I go out later. I could go on with these little vignettes, and a give a few others in the Foreword to this book, but I am sure by now you get the idea. Thus, while we do not instantly log all these observations in our minds as a sort of formal 'cause–effect', in the sense that dark clouds may mean rain, we do so almost unconsciously. But it doesn't stop there and, beyond an avoidance of getting wet or cold, these visual indicators can be important for our well-being and safety. Think about the clues you look for when going to a city for the first time and have a strong desire to avoid being in a dangerous area of a city where you may be vulnerable to robbery. What clues (indicators) do you look for to avoid such a fate? You could opt for an expensive hotel with four or five stars on the reasonable premise that these are unlikely to be in 'bad' areas. The number of stars is intended to be a visual and easily understood indicator of the quality of the hotel, with a typical range in many countries from one star (lowest quality) to five stars (highest quality). The specific criteria used to assign the number of stars may vary between countries, but the basic idea remains – more stars are better. Others may look more widely than just the quality of the hotel and consult sources available via the internet on crime statistics or perhaps just look to see what newspapers and 'blogs' say regarding violent crime and robbery. Other visual clues while physically in the area could be the extent of threatening graffiti on walls, such as gang signs or, at an extreme, the presence of real 'put-offs' such as burnt-out cars in streets, boarded-up shops and the remnants of barricades! These are hardly reassuring signals for those of a nervous disposition and I for one would not feel comfortable. These signs are not infallible, of course, and we need to avoid falling into the trap of thinking that indicators of poverty are the same as those for crime, as they are not. Just because an area might have burnt-out cars and boarded-up shops and houses, it does not mean that all who live there are robbers and bad people! Quite the opposite could be the case.

The above can be regarded almost as a lay view of what an indicator is: A technical sounding term for what can, to be frank, almost be seen as common sense. There is no magic here – indicators are things we look for all of the time to help us make sense of the world, even if they may be for the most part visual clues. Indeed, perhaps clue or signal may be a better way of putting it rather than indicator, and we process so many of them in a single day that we do not even think about it – we just do it. Indeed, they are not only as ubiquitous as love but, and here I will be provocative, they are probably more vital for our day-to-day well-being. This is not me being unromantic but realistic. Benjamin Franklin said, in a letter to the French scientist Jean-Baptiste Leroy, in 1789 following the drafting of the American constitution in 1787:

> Our new Constitution is now established, everything seems to promise it will be durable; but, in this world, nothing is certain except death and taxes.
>
> *(Full text of the letter is available at the Benjamin*
> *Franklin paper repository: franklinpapers.org)*

This is a translation from the original French – Franklin was fluent in that language – but the last few words are often quoted in the English-speaking world even if the letter may not necessarily be the first use of the phrase. The first part of the letter has, I would argue, a point of almost equal relevance for indicators.

It must be remembered that the letter was written by Franklin to Leroy at the time of the French Revolution:

> Are you still living? Or has the mob of Paris mistaken the head of a monop-olizer of knowledge, for a monopolizer of corn, and paraded it about the streets upon a pole.
>
> *(Ibid.)*

Just what the indicators are that the mob would look for in making such a distinc-tion between monopolising heads of knowledge and corn, one can only wonder? Herein rests a lesson that will be returned to throughout this book – relationships between indicators and reality can be imagined.

I will go further than Franklin and say that indicators should be added to this 'death and taxes' duopoly although, as with my suggestion for modifying the title of the Troggs' song, it would not roll off the tongue so well.

> [I]n this world, nothing is certain except death, taxes and indicators.

'Indicator' may not be a word that resonates with most of us in the same way as death and taxes – or perhaps even love – but who knows how such things can catch on in the new world of tweets.

If indicators are such an important aspect of our lives, then why does the term sound so technical? My use of comparable words, such as clues and signals, may well have a more intuitive feel than 'indicators', and when asked what an indica-tor is then most people would probably say that it sounds something 'mathemati-cal' or scientific. That at least is what my students often say when I ask them to define what they think an indicator is: Terms such as numerical, technical, maths, graphs and 'hard' often come up. Feedback from some even suggests that indicators are to be avoided if possible. When asked to elaborate I often hear from students that indicators are numbers listed in long, impenetrable and bor-ing, at least to a non-expert, tables or, worse still, graphs. Rather disconcertingly I have even seen looks of horror in the faces of some students when I tell them that the lecture will be about indicators and that they will be examined on its content. Memorably, two students looked especially horrified in one lecture and they came up at the end and told me that they 'do not do numbers'! But these sentiments from students, whether combined with horror or not, are probably representative of what most would say when confronted with the term.

But the irony is that the common-sense examples I have provided above are not numerical –or are they? Whether we like it or not everything can be given a number – even love. The greyness of the sky can be given a score out of 10 (with 0 perhaps being a perfect blue sky and 10 equating to black clouds), and we can count the number of people wearing coats and express that as a proportion of all those we see, and this can give us a sense of how cold it is. We can even score how threatening an area might seem to us based upon what we see and hear – again perhaps out of 10 with one extreme (10) being 'war zone' and 1 being idyllic. Nonetheless it is true that among indicator technocrats these devices have a meaning that is much more exact than what they would regard, probably with some horror of their own, as my loose wording above. I have used the term in a generalised sense of unconscious and conscious signals and signs that we look for to help us make sense of the world, but the indicator technocrats would see them very much as creations, manufactured tools, designed to help capture or represent the complexity of information. In that context indicators are seen primarily as communication tools – conveying what can be highly complex information collected by one group of people to another group who have to make use of it. Indicators bridge the divide between those whose job it is to know about the world and those who are asked to manage that world. These can, of course, be the same people but usually they are not. However, the skill sets for those who have to 'know' and those who have to 'do' can be very different. For example, economists are specialists and we expect them to know about how economies work. We might quibble about how well they do this given the unpredicted, by most economists anyway, economic crash after 2008, but economies are highly complex entities. To study such entities, economists have to collect a lot of information and try to make sense of it all. This is no easy task as supply and demand within economies can be influenced by the vagaries of politics, and indeed we live in a globalised world where nationalism can trump (pardon the pun) rationality. But then I suppose it is a case of what we mean by rational and whose rationality matters the most. You can also readily assume that such analyses will be complex, and it would take another person versed in economics to really make sense of it all. Rather than hand over all this information and insight verbatim, so to speak, to those who are expected to make rational decisions with it, including politicians, it makes sense for economists to summarise it in the form of a few easily understood indicators. Thus, the politicians, many of whom will not be economists, can use the indicators as the bases for their decisions rather than take on the daunting task of having to read and understand all of the data and analyses. It is akin to us using the star system for hotels reducing the need for customers to be presented with long reams of information about the quality of the chefs, choices available in menus, room service, how often sheets are changed, soaps replenished, and so on. There is obviously a scale difference here in that the management of the economy of a nation has far greater implications than your choice of hotel, although I accept that if you make the wrong choice it might not seem like it; I have been there and can personally vouch for this. But the idea is fundamentally the same.

Maybe it sounds rather tolerant to suggest that indicators are needed to bridge a divide between 'creators' and 'users'. After all, politicians are paid a lot of money by us to govern, and surely we should have high expectations of their expertise? Why can't they get to grips with all this technical knowledge rather than take an easy way out and use indicators created for them? Can we really trust those making these indicators to do it right given that they seem to have so much power? These are important questions, and we will soon get to them, but given that our knowledge has grown rapidly in some fields (including economics) and the methods employed have become ever more complex it does seem rather unfair to expect non-specialists to fully grasp it all. Even politicians need the support of a civil service through which they can govern, and even civil servants need to find ways of accessing the evidence that economists and other specialists can generate. One of the consequences of our increasing knowledge of the world and our impacts upon it is the challenge of making sense of it all so that we can make the 'right' decisions.

So, indicators are very useful things and we use them all of the time even if we are not aware of it. This 'use' of indicators spans the apparently mundane – such as our choice of clothing when leaving the home – to the informing of decisions on national and indeed international scales that affect the lives of millions if not billions. They are inevitable and important, but they are not Laws of Nature. They are human-based interpretations of our world, and to some extent this is influenced by our experiences and our biases. This might sound obvious but, at least in my view, it is a point all too easily forgotten. We may interpret and hence use visual clues in different ways: One person may be alarmed at an area strewn with graffiti and seek to get out of there as fast as possible while another may see this as art and as something to be welcomed. Even economists disagree among themselves, or so we are often told, so is it reasonable to expect them to arrive at a universal set of indicators to promote to users? There is a contrast here between indicators as primarily social constructs and science. After all, the Earth would continue to orbit the Sun even if humans, or anything else for that matter, were not here to see it. That relationship is based upon Isaac Newton's Laws of Motion:

1. Objects at rest or in motion at a particular velocity stay that way unless acted upon by a force.
2. Acceleration of an object is proportional to the force placed on it and the object's mass. The greater the force then the greater the acceleration and larger objects require more force to make them accelerate at the same rate.
3. When two bodies interact, such as striking each other, they do so with the same force. Thus, every force has an equal and opposite reaction.

These laws were articulated by a human being, of course, albeit one of great intellect, but it is likely that their equivalent exists where ever there is intelligent life to discover them. The language used to express the laws may be different, but we can confidently assume that wherever that intelligence may live in the universe these laws will exist. As far as we know they are universal in nature. If ever we make contact with an alien intelligence, it is certain, not just likely, that they will have their own version of these three laws. They really are that important. They apply every time you walk down the street and they help explain the motion of distant galaxies. While they were written by a human being what they describe is not a human construction of reality but a Law of Nature. The words that set out the Laws can be changed, and we can even use diagrams and mathematical symbols to explain what we mean, but the essence of what they articulate remains the same. Whether we like it or not those distant galaxies still exist and orbit each other in ways that can be understood using methods based upon these laws. What we think about the desirability, or not, of these Laws of motion is entirely irrelevant. A politician might not like them and call them 'fake' but, frankly, that does not matter; calling these laws 'fake', no matter how loudly or how often, does not mean they will go away. It is true that science as a means of discovering new knowledge has to take place within a complex and diverse world of values, beliefs and ethics, all of which can influence factors such as available funding and priorities. Our history is replete with science generating knowledge that some do not like, and indeed humans have used knowledge to create harm. After all, knowledge of atoms eventually led to the nuclear bomb. But the problems are often with the application of knowledge – what we do with it – rather than the knowledge itself.

Indicators are fundamentally different to this world of science. Indicators are by their very nature human constructs and based upon assumptions that we make about the world in which we live. Opinions, and the choices that follow, play a very large part indeed. Thus, while dark clouds may suggest that rain is coming, it is not inevitable that it will come. It may be likely but that is not the same as inevitable. Graffiti may be a sign of a bad area of town, but some graffiti is now regarded as art and some of its creators, the artist called Banksy for example, have become famous. Even the star system adopted for hotels might seem like an objective exercise but in fact it is highly dependent not only upon which criteria are included in the system but also how they are assessed and when. Look at a guide book from some years ago and the number of stars allocated for a hotel might be quite different to what they have now. It was one of my mistakes to forget this! Indeed, there are various systems for allocating the number of stars to hotels and restaurants and five stars in one country may not necessarily be the same as five stars in another. Indicators are extensions of us, with all of our values, biases, strengths and failings. The upshot of this is that while a politician cannot change the Laws of Motion, they can, and often do, dispute an indicator and the reasons may or may not be valid.

However, I accept that this apparent divide between Laws of Nature and indicators is not as straightforward as it may first seem. I have mentioned the reality that science as a way of discovering new knowledge takes place within a societal context. Society dictates the science carried out largely because it is taxpayers who pay for it, and society also dictates what uses the knowledge is put to. Hence, science as a process is inevitably selective and scientists as the group who 'do' science have to respond to that selectivity if they want to do science. If they do not respond then the work that they wish to do may not be done by them, although someone else, at some time, in some other place, may have that opportunity. But whoever does the science, when they do it or where they are, the new knowledge that they find is resilient; by definition, with the right resources anyone can repeat what they have done and arrive at the same results. Indicators, by way of contrast, are transient tools that are highly subjective. Anyone can make an indicator, and the chances are that your indicators in any one situation may not be the same as mine. Indeed, yours may be better than mine. My visual clues for what I perceive as a rough area of town may not be shared by others, and I may be wrong in my assessment. But while an intelligent alien species advanced enough to understand the motion of planets, stars and galaxies will certainly have the equivalent of our Laws of Motion they are almost just as certain to have indicators as well. They too will have to live by trying to make sense of their world and no doubt some will specialise in some aspects of knowledge generation that they will need to share with others who are not so well versed in the specialisation. Whether they have hotels is another matter.

So why have I written a book about indicators? I feel it is important for us to know that indicators matter, and we need to know the benefits and pitfalls that surround them. This is true at a personal level, although I have no intention of writing a book on common sense – indeed whose 'common sense' would apply? Instead, my target is those indicators that operate at a larger scale: The ones that equate with those hard, numerical, tabulated and, at least for some, boring devices. This sounds like a challenge and indeed it is, but these more technocratic indicators also matter to all of us, although the relationship can often be far less apparent within our everyday lives. While the star system for hotels is an indicator that many of us will encounter at some stage in our lives, even if only on holiday, the other indicators I will discuss in this book, such as Gross Domestic Product (GDP) in economics, are perhaps not so obviously useful or relevant for most of us. But decisions taken over the economy can impact upon us all, as many of us know all too well following the economic slump after 2008, but this is by no means the only example. Indeed, this book will present a number of indicators, or more precisely 'indices', where an index is a combination of indicators, showing what countries and indeed the world look like when using these numbers. It is almost as if they act as different colour filters for seeing our planet: Use a different filter and the world looks different. What fascinates me about indicators is the very fact that they are not science but are extensions of ourselves. Looking at the world with these indicators is like looking at the world in the many ways that people think are important. What do these indicators tell us about the world but, just as important, what do they tell us about those who have created them? Answers to these questions may well surprise you.

Countries compared: The Olympic dream

In order to set the scene for my visualisation of the world in numbers I will begin with a table of nation states that is probably familiar to almost everyone: The final table of medals following the 2012 Olympic Games in London. Now

I must first declare any potential bias at this point as I know from experience that this can be something of a hot topic. People can be very passionate about the Olympics and I do not want to start off on the wrong foot. I have selected the 2012 medal table rather than the more recent 2016 Olympic medal table for the Games held in Rio de Janeiro, Brazil, for a number of reasons.[1] First, I am British by birth and I live in West London, and thus was affected by the Games even though they were held in East London. I did not attend a single Olympic event although I did witness the torch coming through my part of the city a few days before the opening ceremony. I also did not see much of the Games on television, with the exception of a couple of the high-profile events such as the Men's 100 m sprint won by Usain Bolt of Jamaica who set a new Olympic record of 9.63 seconds. I also did not see much of the Rio Games although I was far less personally affected by them. Second, there is always the claim of national bias to contend with. The UK came third in the 2012 medal table, behind China and the USA, but came second in the 2016 table. Let it not be said that I am selectively opting for a table where my own country came second in the world! I certainly cannot claim by any stretch of the imagination to be a fan of the Olympics, although I also have nothing at all against them and will readily express my admiration for all those athletes who took part. I am an Olympic agnostic and my choice of the Olympic table is not driven by any fanaticism for the occasion or any desire to highlight how well or poorly some countries did, including my own, relative to others. It serves my purpose purely because it is an international comparison table that many readers will be aware of and will probably have been seen on a regular basis while the Olympics were taking place.

It should also be noted that I have not included the medals from the Para-Olympic Games, also held in London following the Olympic Games. The Para-Olympics is a more recent invention, dating back to 1948, and is the same as the Olympics but the contestants have some form of disability. I have often seen complaints that the Para-Olympics has a much lower profile in the media than the Olympics, and from the little I know about the subject and the coverage I witnessed in the media, at least in the UK, this does seem like a reasonable criticism. I certainly have no intention of getting involved in that debate, but given that all I am trying to do is highlight the sort of between-country comparisons that often occur with indicators, and these can be complicated enough as it is, I have no wish to confuse the reader by using two sets of tables for different types of Olympic games.

With all that off my chest, we can ask some questions about how the medal table was created and what were the assumptions behind it? Why these questions matter will be covered later on, but first we need to look at the final medal table for the 2012 Olympics, which is shown in Table 1.1.

It is something of a long list, as almost every country in the world sent a delegation to the Games, with lots of zeros – but I make no excuses for that. Every country that took part deserves a mention, even if its team did not win a single medal. A couple of points are worth noting for those who know as little as I do about the Olympics or perhaps even less (if possible). First, of course, it

TABLE 1.1 Final medal table for the 2012 Olympic Games held in London (this ranking follows the rules commonly applied with Olympic medal tables whereby gold takes precedence over silver and silver takes precedence over bronze)

Country	Medals won			
	Gold	Silver	Bronze	Total
United States of America	46	29	29	104
China, People's Republic of	38	27	23	88
Great Britain	29	17	19	65
Russia	24	26	32	82
Korea, Republic of	13	8	7	28
Germany	11	19	14	44
France	11	11	12	34
Italy	8	9	11	28
Hungary	8	4	5	17
Australia	7	16	12	35
Japan	7	4	17	28
Kazakhstan	7	1	5	13
Netherlands	6	6	8	20
Ukraine	6	5	9	20
New Zealand	6	2	5	13
Cuba	5	3	6	14
Iran	4	5	3	12
Jamaica	4	4	4	12
Czech Republic	4	3	3	10
DPR Korea	4	0	2	6
Spain	3	10	4	17
Brazil	3	5	9	17
South Africa	3	2	1	6
Ethiopia	3	1	3	7
Croatia	3	1	2	6
Belarus	2	5	5	12
Romania	2	5	2	9
Kenya	2	4	5	11
Denmark	2	4	3	9
Azerbaijan	2	2	6	10
Poland	2	2	6	10
Turkey	2	2	1	5
Switzerland	2	2	0	4
Lithuania	2	1	2	5
Norway	2	1	1	4
Canada	1	5	12	18
Sweden	1	4	3	8
Colombia	1	3	4	8
Georgia	1	3	3	7
Mexico	1	3	3	7
Ireland	1	1	3	5
Argentina	1	1	2	4

(Continued)

TABLE 1.1 Continued

Country	Medals won			
	Gold	Silver	Bronze	Total
Serbia	1	1	2	4
Slovenia	1	1	2	4
Tunisia	1	1	1	3
Dominican Republic	1	1	0	2
Trinidad and Tobago	1	0	3	4
Uzbekistan	1	0	3	4
Latvia	1	0	1	2
Algeria	1	0	0	1
Bahamas	1	0	0	1
Grenada	1	0	0	1
Uganda	1	0	0	1
Venezuela	1	0	0	1
India	0	2	4	6
Mongolia	0	2	3	5
Thailand	0	2	1	3
Egypt	0	2	0	2
Slovakia	0	1	3	4
Armenia	0	1	2	3
Belgium	0	1	2	3
Finland	0	1	2	3
Bulgaria	0	1	1	2
Estonia	0	1	1	2
Indonesia	0	1	1	2
Malaysia	0	1	1	2
Puerto Rico	0	1	1	2
Taiwan (2)	0	1	1	2
Botswana	0	1	0	1
Cyprus	0	1	0	1
Gabon	0	1	0	1
Guatemala	0	1	0	1
Montenegro	0	1	0	1
Portugal	0	1	0	1
Greece	0	0	2	2
Moldova	0	0	2	2
Qatar	0	0	2	2
Singapore	0	0	2	2
Afghanistan	0	0	1	1
Bahrain	0	0	1	1
Hong Kong	0	0	1	1
Kuwait	0	0	1	1
Morocco	0	0	1	1
Saudi Arabia	0	0	1	1
Tajikistan	0	0	1	1
Albania	0	0	0	0
American Virgin Islands	0	0	0	0

Andorra	0	0	0	0
Angola	0	0	0	0
Antigua and Barbuda	0	0	0	0
Aruba	0	0	0	0
Austria	0	0	0	0
Bangladesh	0	0	0	0
Barbados	0	0	0	0
Belize	0	0	0	0
Benin	0	0	0	0
Bermuda	0	0	0	0
Bhutan	0	0	0	0
Bolivia	0	0	0	0
Bosnia and Herzegovina	0	0	0	0
British Virgin Islands	0	0	0	0
Brunei Darussalam	0	0	0	0
Burkina Faso	0	0	0	0
Burma (Myanmar)	0	0	0	0
Burundi	0	0	0	0
Cambodia	0	0	0	0
Cameroon	0	0	0	0
Cape Verde	0	0	0	0
Cayman Islands	0	0	0	0
Central African Republic	0	0	0	0
Chad	0	0	0	0
Chile	0	0	0	0
Comoros	0	0	0	0
Congo	0	0	0	0
Congo, the Democratic Republic of the	0	0	0	0
Cook Islands	0	0	0	0
Costa Rica	0	0	0	0
Côte d'Ivoire	0	0	0	0
Djibouti	0	0	0	0
Dominica	0	0	0	0
Ecuador	0	0	0	0
El Salvador	0	0	0	0
Equatorial Guinea	0	0	0	0
Eritrea	0	0	0	0
Fiji	0	0	0	0
Former Yugoslav Republic of Macedonia	0	0	0	0
Gambia	0	0	0	0
Gaza Strip/Palestine/West Bank	0	0	0	0
Ghana	0	0	0	0
Guam	0	0	0	0
Guinea	0	0	0	0
Guinea-Bissau	0	0	0	0
Guyana	0	0	0	0
Haiti	0	0	0	0
Honduras	0	0	0	0
Iceland	0	0	0	0

(Continued)

TABLE 1.1 Continued

Country	Medals won			
	Gold	Silver	Bronze	Total
Iraq	0	0	0	0
Israel	0	0	0	0
Jordan	0	0	0	0
Kiribati	0	0	0	0
Kyrgyzstan	0	0	0	0
Laos	0	0	0	0
Lebanon	0	0	0	0
Lesotho	0	0	0	0
Liberia	0	0	0	0
Libya	0	0	0	0
Liechtenstein	0	0	0	0
Luxembourg	0	0	0	0
Madagascar	0	0	0	0
Malawi	0	0	0	0
Maldives	0	0	0	0
Mali	0	0	0	0
Malta	0	0	0	0
Marshall Islands	0	0	0	0
Mauritania	0	0	0	0
Mauritius	0	0	0	0
Micronesia	0	0	0	0
Monaco	0	0	0	0
Mozambique	0	0	0	0
Namibia	0	0	0	0
Nauru	0	0	0	0
Nepal	0	0	0	0
Nicaragua	0	0	0	0
Niger	0	0	0	0
Nigeria	0	0	0	0
Oman	0	0	0	0
Pakistan	0	0	0	0
Palau	0	0	0	0
Panama	0	0	0	0
Papua New Guinea	0	0	0	0
Paraguay	0	0	0	0
Peru	0	0	0	0
Philippines	0	0	0	0
Rwanda	0	0	0	0
Samoa	0	0	0	0
Samoa, American	0	0	0	0
San Marino	0	0	0	0
Sao Tome and Principe	0	0	0	0
Senegal	0	0	0	0
Seychelles	0	0	0	0
Sierra Leone	0	0	0	0

Solomon Islands	0	0	0	0
Somalia	0	0	0	0
Sri Lanka	0	0	0	0
St Kitts and Nevis	0	0	0	0
St Lucia	0	0	0	0
St Vincent and the Grenadines	0	0	0	0
Sudan	0	0	0	0
Suriname	0	0	0	0
Swaziland	0	0	0	0
Syria	0	0	0	0
Tanzania	0	0	0	0
Timor-Leste	0	0	0	0
Togo	0	0	0	0
Tonga	0	0	0	0
Turkmenistan	0	0	0	0
Tuvalu	0	0	0	0
United Arab Emirates	0	0	0	0
Uruguay	0	0	0	0
Vanuatu	0	0	0	0
Viet Nam	0	0	0	0
Yemen	0	0	0	0
Zambia	0	0	0	0
Zimbabwe	0	0	0	0

Source of data: https://www.olympic.org/olympic-results.

is a ranking of countries as most competitors in the Olympic Games represent nations and wear clothing of varying hues and styles that makes that allegiance clear. I say 'most' because there are some competitors, although not many (only four in the London Games of 2012), who compete under the Olympic flag rather than a national one. Why this happens need not be covered here, but for the sake of simplicity you will not find 'no country' in Table 1.1. Second, the ranking in this table follows a very simple set of rules. Countries are first ranked by the number of gold medals won by their athletes (those that win events receive a gold medal for doing so) then by the number of silver (those who come second receive a silver medal) and finally by the number of bronze (by now you can work this out for yourself). It is not a ranking based on the total number of medals won by a country (the column on the far right-hand side of the table) and this makes a big difference. Therefore, Great Britain (GB), the host nation, in third place in the table, came above Russia even though it had fewer medals. Russia had 82 medals in total while GB had 65. But GB's 29 gold medals put it above Russia, which had 24 gold medals, even though Russia had more silver and bronze medals. It also explains why Algeria, Bahamas, Grenada, Uganda and Venezuela, all of whom had only one gold medal, are ranked above many countries that had a lot more medals. Indeed, in theory it is possible for a country to have 100 silver and bronze medals, representing a huge achievement by any standard, but be ranked below a country having just one – as long as it is gold. There are other ways in which the ranking could be done but this is the most commonly reported method. Please note that I am not saying that one way of doing the ranking is any better or worse than any other. All I am doing is stating the ranking and assumptions behind it as typically reported. Also note that many countries, nearly half in fact, which took part in the Olympics received no medal at all: They did not come first, second or third at any event. Hence the many zeros in the table. For these countries and indeed any countries that 'tie' in terms of the rules set out above (and this is common when the number of medals won is quite low), the ranking is purely alphabetical. That is why Zimbabwe is at the bottom of the table. Thus, in terms of the table it pays to have a country name that starts with 'A' – as with 'Albania', which had no medals but it is higher up the table than Zimbabwe! Indeed, it might surprise you to learn that just over half of the total number of medals in those Games were won by the top ten countries in the table; barely 5% of the countries in the table won 55 per cent of the medals and nearly 60% of the countries in the table (119 of them) won no medal at all. The table has far more zeros in it that any other number. It is also worth noting that the winning margin of an event is not included in any shape or form. Hence, a very close victory in an event, perhaps by a split second in the case of races or a single point with events based on judges scoring a performance (as in diving), is enough to win one type of medal over another or indeed to win any medal at all. The winner, or top three, take it all, and these fine margins matter. In theory, a country having 100 silver and bronze medals that were all won emphatically, can be ranked below a country having a single gold medal that was won by the smallest of margins, perhaps a split second. Also, there is no adjustment in terms of the number of events for which a country had athletes entered. A country may only have entered one athlete and won a medal, representing a 100% success rate. This theoretical position did not occur in London 2012, but Botswana only entered 4 athletes and won a silver medal. Other countries only having a silver medal as their 'total' for the Games include Cyprus with 13 competitors, Gabon with 28 competitors, and Portugal with 80 competitors. The success rate of Botswana is certainly higher than those of Cyprus, Gabon and Portugal, but is this all that matters? Portugal may only have won one medal out of its team of 80, but surely one can commend the country's spirit of engagement in the Olympic ideal?

The medal table, or perhaps more accurately the version of it in Table 1.1, received wide coverage during the Olympics via just about every media outlet imaginable. But it was the choice of certain individuals to select the rules upon which it is based and, unlike Newton's Laws of Motion, the rules for the construction of the table are not Laws of Nature. They reflect opinion and nothing more, even if those who created the rules feel they had good reasons for doing so. An alternative would be to rank countries simply in terms of the number of medals each country won. Thus, a bronze medal would have an equal weighting to a gold medal. The results based on these alternative rules for those countries that won at least one medal (just to keep the table as short as possible), was shown in Table 1.2.

TABLE 1.2 Revised 2012 Olympic Games medal table based solely on the number of medals

Country	Medals won			
	Gold	Silver	Bronze	Total
United States of America	46	29	29	104
China, People's Republic of	38	27	23	88
Russia	24	26	32	82
Great Britain	29	17	19	65
Germany	11	19	14	44
Australia	7	16	12	35
France	11	11	12	34
Italy	8	9	11	28
Japan	7	4	17	28
Korea, Republic of	13	8	7	28
Netherlands	6	6	8	20
Ukraine	6	5	9	20
Canada	1	5	12	18
Brazil	3	5	9	17
Hungary	8	4	5	17
Spain	3	10	4	17
Cuba	5	3	6	14
Kazakhstan	7	1	5	13
New Zealand	6	2	5	13
Belarus	2	5	5	12
Iran	4	5	3	12
Jamaica	4	4	4	12
Kenya	2	4	5	11
Azerbaijan	2	2	6	10
Czech Republic	4	3	3	10
Poland	2	2	6	10
Denmark	2	4	3	9
Romania	2	5	2	9
Colombia	1	3	4	8
Sweden	1	4	3	8
Ethiopia	3	1	3	7
Georgia	1	3	3	7
Mexico	1	3	3	7
Croatia	3	1	2	6
DPR Korea	4	0	2	6
India	0	2	4	6
South Africa	3	2	1	6
Ireland	1	1	3	5
Lithuania	2	1	2	5
Mongolia	0	2	3	5
Turkey	2	2	1	5
Argentina	1	1	2	4
Norway	2	1	1	4

(Continued)

TABLE 1.2 Continued

Country	Medals won			
	Gold	Silver	Bronze	Total
Serbia	1	1	2	4
Slovakia	0	1	3	4
Slovenia	1	1	2	4
Switzerland	2	2	0	4
Trinidad and Tobago	1	0	3	4
Uzbekistan	1	0	3	4
Armenia	0	1	2	3
Belgium	0	1	2	3
Finland	0	1	2	3
Thailand	0	2	1	3
Tunisia	1	1	1	3
Bulgaria	0	1	1	2
Dominican Republic	1	1	0	2
Egypt	0	2	0	2
Estonia	0	1	1	2
Greece	0	0	2	2
Indonesia	0	1	1	2
Latvia	1	0	1	2
Malaysia	0	1	1	2
Moldova	0	0	2	2
Puerto Rico	0	1	1	2
Qatar	0	0	2	2
Singapore	0	0	2	2
Taiwan	0	1	1	2
Afghanistan	0	0	1	1
Algeria	1	0	0	1
Bahamas	1	0	0	1
Bahrain	0	0	1	1
Botswana	0	1	0	1
Cyprus	0	1	0	1
Gabon	0	1	0	1
Grenada	1	0	0	1
Guatemala	0	1	0	1
Hong Kong	0	0	1	1
Kuwait	0	0	1	1
Montenegro	0	1	0	1
Morocco	0	0	1	1
Portugal	0	1	0	1
Saudi Arabia	0	0	1	1
Tajikistan	0	0	1	1
Uganda	1	0	0	1
Venezuela	1	0	0	1

Medal counts are the same as for Table 1.1 but in this case the ranking of each country in the table is based solely upon the number of medals that each country was successful in winning. Countries not winning any medals have been omitted.

Source of data: https://www.olympic.org/olympic-results.

The alphabetical ranking for countries that tie has been maintained in Table 1.2. Some countries have done much better in this new version of the medal table, while others have slipped down. My country, GB, dropped from third to fourth. But this would seem to be rather unfair. Surely common sense tells us that a gold medal should be of greater 'value' (and I am not talking about financial value) than a bronze medal? Both Uganda and Venezuela have single medals, but they are gold – yet they appear at the bottom of the table because their names start with 'U' and 'V' respectively. Hence the original ranking has some virtue in the sense that it takes the number of gold medals as its starting point and carries on from there. But can that be said to be fair? What about a country like India that won six medals in total, two of which were silver and four that were bronze? India ranks below a number of countries that won just one gold medal, namely Grenada, Bahamas, Venezuela, Uganda and Algeria. Should gold medals be rated so highly that they appear to trump all the others? Well the answer is a clear 'yes' in the original form of the table but to some, including me, it does seem a bit unjust. A compromise between these two variants of the medal table might be to allocate 'weightings' to the medals. Thus, gold could attract a weighting of three while silver attracts two and bronze one. The table generated in this case is presented as Table 1.3, while again only including those countries that won at least one medal and keeping the alphabetical rule in place for ties.

The shift in rank is subtle and hardly noticeable towards the top end of the table, but it does make a difference to some countries. Uganda and Venezuela are no longer at the foot of the table, for example. We can play around with this to our hearts content. Rather than use 3-2-1 for the relative weighting of gold-silver-bronze we could use 4-2-1 on the grounds that each medal is worth twice the one below it and so on. The sky really is the limit here and I will not bore the reader with presenting any more permutations arising from such different weightings. Needless to say, such changes in assumption can make a difference to the country ranking.

But what exactly is this table a reflection of? What probably comes immediately to mind is that the table is a reflection of a country's ability to get its athletes to the level where they can successfully compete against others and become among the best three in the world at their chosen event. In other words, the ranking equates to sporting success. This all sounds rather obvious, of course, and it is perhaps not surprising that a high ranking in the medal league table is a source of national pride and gives a great sense of achievement. In the UK the Olympic team were understandably called 'heroes' in the national press for their achievement in winning 65 medals and coming third in the 'original' form of the medal table. Some of the athletes were given well-deserved knighthoods along with other awards. The fact that it was a 'home' Games made it all the more exciting and heightened that sense of pride. But is ranking in the medal table really a reflection of sporting prowess at the national scale? Well partly. The number of medals that a country has a chance of winning will obviously reflect the number of events for which that country has entered its athletes, although this is not necessarily the same thing as the size

TABLE 1.3 The 2012 Olympic Games medal table based on a scoring system for medals

Country	Medals won			
	Gold	Silver	Bronze	Total score
United States of America	46	29	29	225
China, People's Republic of	38	27	23	191
Russia	24	26	32	156
Great Britain	29	17	19	140
Germany	11	19	14	85
France	11	11	12	67
Australia	7	16	12	65
Korea, Republic of	13	8	7	62
Italy	8	9	11	53
Japan	7	4	17	46
Netherlands	6	6	8	38
Hungary	8	4	5	37
Ukraine	6	5	9	37
Spain	3	10	4	33
Brazil	3	5	9	28
Kazakhstan	7	1	5	28
Cuba	5	3	6	27
New Zealand	6	2	5	27
Canada	1	5	12	25
Iran	4	5	3	25
Jamaica	4	4	4	24
Belarus	2	5	5	21
Czech Republic	4	3	3	21
Kenya	2	4	5	19
Romania	2	5	2	18
Denmark	2	4	3	17
Azerbaijan	2	2	6	16
Poland	2	2	6	16
DPR Korea	4	0	2	14
Ethiopia	3	1	3	14
South Africa	3	2	1	14
Sweden	1	4	3	14
Colombia	1	3	4	13
Croatia	3	1	2	13
Georgia	1	3	3	12
Mexico	1	3	3	12
Turkey	2	2	1	11
Lithuania	2	1	2	10
Switzerland	2	2	0	10
Norway	2	1	1	9
India	0	2	4	8
Ireland	1	1	3	8
Argentina	1	1	2	7
Mongolia	0	2	3	7
Serbia	1	1	2	7

Slovenia	1	1	2	7
Trinidad and Tobago	1	0	3	6
Tunisia	1	1	1	6
Uzbekistan	1	0	3	6
Dominican Republic	1	1	0	5
Slovakia	0	1	3	5
Thailand	0	2	1	5
Armenia	0	1	2	4
Belgium	0	1	2	4
Egypt	0	2	0	4
Finland	0	1	2	4
Latvia	1	0	1	4
Algeria	1	0	0	3
Bahamas	1	0	0	3
Bulgaria	0	1	1	3
Estonia	0	1	1	3
Grenada	1	0	0	3
Indonesia	0	1	1	3
Malaysia	0	1	1	3
Puerto Rico	0	1	1	3
Taiwan	0	1	1	3
Uganda	1	0	0	3
Venezuela	1	0	0	3
Botswana	0	1	0	2
Cyprus	0	1	0	2
Gabon	0	1	0	2
Greece	0	0	2	2
Guatemala	0	1	0	2
Moldova	0	0	2	2
Montenegro	0	1	0	2
Portugal	0	1	0	2
Qatar	0	0	2	2
Singapore	0	0	2	2
Afghanistan	0	0	1	1
Bahrain	0	0	1	1
Hong Kong	0	0	1	1
Kuwait	0	0	1	1
Morocco	0	0	1	1
Saudi Arabia	0	0	1	1
Tajikistan	0	0	1	1

Note: This table is based on the same medal count as Tables 1.1 and 1.2 but here a score has been awarded to each type of medal: Gold = 3 points, Silver = 2 points and Bronze = 1 point. Countries with no medal have been omitted.

Source of data: https://www.olympic.org/olympic-results.

of the Olympic team. A country that enters 11 athletes into the same individual-based event, such as the high jump, can only win 3 medals at most. Indeed, if that event was soccer, a team event, then it can only win one medal. But if those 11 competitors were entered into 11 events then it is possible, although not necessarily likely, to win 11 medals of varying hues. Even a country with a single athlete can be involved in more than one event. Therefore, ranking can be influenced by the number of events a country enters its athletes into and there are no rules that stipulate the minimum or maximum number of events a country can enter. This is, of course, a simplification as quality of athletes is far more important than quantity, but assuming that a country has a reasonable chance of winning the events it has entered for then team size may be a factor. In Table 1.4 I have provided yet another form of the medal table with an adjustment made for the size of the team. All I have done is divide the number of medals by the team size. This is a somewhat crude adjustment, I must admit, as it should really be adjusted by the number of events entered rather than team size, but it helps reinforce the point. The table represents a sort of 'success rate' of a country's athletes.

Table 1.4, which arguably is a fairer representation of medal 'success' at the Olympics, provides a very different ranking to that based upon any of the methods described to date. Top of the pile is Botswana followed by Jamaica and Team GB drops down to 18th position. By the way, the numbers in the table are given here to 3 decimal places but the ranking has been done on the basis of 9 decimal places! This is perhaps a bit extreme, but it does mean the apparent 'ties' in the table were not resolved on the basis of the name of the country but by the full value of the number of medals per athlete, even if the difference was very small. This explains why Thailand, for example, is ranked higher than Lithuania. The message to take away from all of this is that a careful selection of team size and events entered could help enhance the prospects of a country doing well in a medal table based on success rate. But what other alternatives to the original medal table are possible? It was certainly the case that some countries spent a lot of money on facilities preparing for the 2012 Olympics. A report in the *Guardian* newspaper carried the 'costs' associated with each medal (gold, silver and bronze) won by Team GB, and this is summarised as Table 1.5.

The costs associated with each sport include training, equipment and other facilities and can vary dramatically from one sport to another. It is known that the total UK Sport (the UK's 'sports agency') funding for the London 2012 Olympics was £264,143,753, and this money was supplemented by a private sponsorship scheme known as 'Team 2012'. A quarter of a billion pounds sterling from UK Sport is a large sum of money by any standard and may surprise the reader, but then again it may not. Olympic success has been seen as highly desirable by many countries over many years, and those countries have invested resources accordingly. The Olympics used to be a great bastion of amateurism, where athletes are not full-time and train alongside doing their regular job. For many that ideal still holds, but increasingly the athletes are becoming professional and this requires funding.

TABLE 1.4 Medal league table from the 2012 Olympic Games based on success rate of a country's athletes: The number of medals won per team member

Country	Medals/team member
Botswana	0.25
Jamaica	0.24
China, People's Republic of	0.237
Iran	0.226
Kenya	0.22
Ethiopia	0.2
Georgia	0.2
United States of America	0.196
Azerbaijan	0.189
Russia	0.189
Mongolia	0.172
Afghanistan	0.167
Qatar	0.167
Trinidad and Tobago	0.129
Cuba	0.127
Japan	0.125
Armenia	0.12
Great Britain	0.117
Kazakhstan	0.113
Germany	0.111
Netherlands	0.11
Korea, Republic of	0.11
DPR Korea	0.109
Hungary	0.108
France	0.101
Grenada	0.1
Kuwait	0.1
Italy	0.1
Indonesia	0.091
Moldova	0.091
Romania	0.087
Singapore	0.087
Slovakia	0.087
Australia	0.085
Ukraine	0.085
Bahrain	0.083
Thailand	0.081
Lithuania	0.081
Puerto Rico	0.08
Denmark	0.078
Cyprus	0.077
Ireland	0.076
Czech Republic	0.075
Uzbekistan	0.074

(*Continued*)

TABLE 1.4 Continued

Country	Medals/team member
Colombia	0.074
India	0.072
Belarus	0.07
Malaysia	0.067
New Zealand	0.066
Mexico	0.066
Canada	0.065
Brazil	0.064
Tajikistan	0.063
Uganda	0.063
Norway	0.062
Estonia	0.061
Slovenia	0.059
Spain	0.059
Dominican Republic	0.057
Sweden	0.057
Croatia	0.055
Finland	0.054
Guatemala	0.053
Saudi Arabia	0.053
Poland	0.046
Taiwan	0.045
South Africa	0.045
Turkey	0.044
Latvia	0.043
Bahamas	0.038
Switzerland	0.038
Gabon	0.036
Tunisia	0.036
Serbia	0.034
Bulgaria	0.032
Montenegro	0.029
Argentina	0.028
Algeria	0.026
Belgium	0.025
Hong Kong	0.024
Greece	0.019
Egypt	0.017
Venezuela	0.015
Morocco	0.014
Portugal	0.013

Source of data: https://www.olympic.org/olympic-results.

TABLE 1.5 Cost (£) per medal won by the team from Great Britain in the 2012 London Olympic Games

Sport	Cost per medal won (£)
Hockey	15,013,200
Swimming	8,381,533
Diving	6,535,700
Modern pentathlon	6,288,800
Sailing	4,588,540
Athletics	4,191,333
Canoeing	4,044,175
Judo	3,749,000
Rowing	3,031,956
Gymnastics	2,692,650
Equestrian	2,679,020
Triathlon	2,645,650
Shooting	2,461,866
Taekwondo	2,416,800
Cycling	2,169,333
Boxing	1,910,280

Source: The data used to construct this table are available from: www.theguardian.com/sport/datablog/2012/aug/13/olympics-2012-cost-per-medal-team-gb-funding#data.

Hence it does not seem unreasonable to assume that the number of medals a team wins will be influenced in part by the money spent on that team, but here the rationale starts to get rather fuzzy. First, there is the assumption that more money invested in facilities and athletes results in better performance. A lot of money may be spent but this may not necessarily mean more medals. Second, it is not all that easy to get the figures for such investment for every country taking part. Indeed, what investment should we include here? Do we just focus on money that may be coming from government or do we include sponsorship deals that some athletes (certainly not all) may have attracted? This is something of a can of worms, although various people have tried to relate a country's performance to investment, more often or not they had to approximate this in quite crude ways. For example, some have used the economic wealth of a country as a sort of indicator for investment into its athletes. This is a huge leap of faith for all the reasons given above, but I suppose the view taken is that it is better to make some adjustment for wealth and, by loose association, investment, rather than make no adjustment at all. Figures are readily available from many international bodies to help gauge a country's economic wealth, although as we shall see later this is also open to all sorts of problems with measurement and interpretation. Unsurprisingly, well to me at any rate, the attempts I have seen to link Olympic performance to national wealth have not told us very much other than it may have been a factor in some cases but not in others. Not a conclusion that is all that helpful.

If performance in the Olympic medal table can be influenced by a number of factors, then how should a country respond to this if it wants to do better next time around? Well I suppose the obvious answer is 'all of the above': Increase the number of events entered and enhance the quality of your athletes by better selection and more investment. All sounds very obvious really, although it is far from being easy or indeed cheap to achieve. What about simply asking for a change in medal table methodology that gives you the best ranking possible? Botswana, for example, could argue for the version of the table based on success rate (Table 1.4) that eliminates factors such as team size that in turn could be linked to wealth. There is certainly a compelling logic here, and a change in league table methodology is by far the easiest and cheapest option for Botswana to go for, and can hardly be called cheating if they can convince all those involved in compiling and publishing such things that their suggestion is the best. However, there is the problem that many other countries will not agree. The 'winners' in the original form of the league table (Table 1.1) are more likely to argue that it should be maintained. There is nothing preventing any country from juggling the league table methodology to best enhance its own standing, but what matters is getting acceptance from all those involved.

But aside from all this tortuous manipulation of rankings we should ask ourselves what the point of the table is in the first place. After all, it is a relatively recent invention given the history of the Olympics and it is not an 'official' device used by anyone to make decisions about who should get to host the Games in the future. There are no monetary rewards for doing well in the medal table, although individual competitors could do well in terms of sponsorship, media appearances and so on. It is, of course, a matter of national pride and, in a sense, it is a device for translating what are individual or team performances across a host of entirely unrelated sports so as to allow a comparison between nations. The table does feature as a device in the setting of targets by national sporting bodies. For example, we have the following quote from the Chief Executive of UK Sport, Liz Nicholl, regarding targets for the 2012 Olympics in London:

> At least 48 medals is a realistic target in at least 12 sports and top four on the medal table ... We aspire for them to be more than 48 medals in more than 12 sports ... We can't be more specific than that.
>
> (The Telegraph, 2012)

The good news for Ms Nicholl and indeed her country is that using most of the rankings given here then Team GB met its target both in terms of the number of models and a top four place. The only ranking in the tables that I have provided where Team GB failed in its top four position was in the table based on medals won per athlete.

Life beyond the Olympics

The Olympic medal table is an example of the world in numbers. Each nation state ends up with a single numerical value: Its rank in the medal league table. All of the effort, tears, planning, training, investment, publicity, hype and politics over many years prior to the Games becomes condensed into a single number. But that number matters to a lot of people. I need to be careful here because while some may see sport as nothing more than absorbing viewing on television, for others it is a valuable way of keeping fit and indeed the success of one's team can be an important contribution to one's sense of well-being. While I cannot claim to be an avid fan of the Olympics I do follow the fortunes of Chelsea Football Club here in Britain and watch as many of their games as I can. I like to see them win, of course, and at the time of writing they have been doing that a lot. This takes us to a different league table, of course, but I will not go there in this book. Dissecting one sports-based table is enough. But while sport does matter for many it is not in the same 'league' (pardon the pun) as addressing important issues such as the poverty and inequality that still exist in many parts of the world. In the early years of the twenty-first century, many people still do not have access to the basics of life – water, food and even housing – and abject poverty can exist alongside great wealth, perhaps a matter of just a few metres away, Even in the so-called developed countries of the world we can still see deprivation existing next to riches. Nor does sport equate in most people's minds with the importance of being able to keep their job, or to be able to educate their children, or to be able to live in surroundings that are safe. Bill Shankly, a famous football manager in Britain , once of Liverpool FC, said on a TV chat show in 1964 that:

Football's not a matter of life and death ... it's more important than that.

He was speaking very much with his tongue in his cheek: Sport is obviously not life or death for the vast majority of people who live on this planet. Many of them were not able to follow the London Olympics on TV or radio, not because they may have been uninterested but because they simply could not afford to. Their priorities were and still are elsewhere.

What is the link between the Olympic medal table and the theme of this book? Well this book also seeks to set out some of the ways in which we view the world today using indicators. We are used to seeing pictures of our planet as that predominantly blue disk against the blackness of space and it is a beautiful sight. The pictures *Earth Rise*, taken by astronauts on the Apollo missions as they orbited the moon, are burnt into the psyche of many of my generation who went on to work on environmental issues. So too are the pictures from the distant *Voyager* spacecraft taken when looking back towards the Earth, which is nothing more than a blue dot of a few pixels in a deep black background. Such pictures help give us a sense of perspective. They give us that sense of isolation in the blackness of space and a realisation that this is our home and we have nowhere else to go that provides the same resources we need to live. We may well develop the technology that allows us to explore and eventually live on other planets, but this planet will always be the home of the human race. The world is diverse. Genetically the human race is remarkably homogeneous compared with other animals, largely, it is thought, because we almost became extinct some 150,000 years ago and those few thousand individuals that survived have provided us with the level of genetic diversity we see today. Thankfully we have made up for that genetic narrowness with a blossoming of diversity in culture, language, art and so on. But there is diversity and diversity. Diversity in culture, ways of living and language is one thing, but when one end of that diversity is defined by extreme poverty, insecurity and fear then that is something else. Assessing such diversity so that people can address it is not an easy task, but do we need to measure? Peter Drucker, the man who has been said to have invented modern business management, has often been attributed with the phrase:

> If you can't measure it, you can't improve it.

These words are often repeated in a variety of disciplines, not just business, and I have heard them expressed during numerous lectures by indicator people to justify the need for their craft. But the sentiment is debatable. Indeed, Bill Hennessy, also a business consultant, makes a convincing case that Drucker never said or claimed any such thing:

> This is the most evil and destructive Drucker quote of all time. I hear it at least once a month, usually to justify elimination of tasks that cannot easily be measured using the kinds of simple yardsticks executives fancy. Ya know, unmeasurable work like ingenuity, coaching, innovation, creativity, and, Drucker's favorite, imagination.
>
> *(Available at http://billhennessy.com/simple-strategies/*
> *2015/09/09/i-wish-drucker-never-said-it)*

Another oft-cited quote that says the opposite of the claimed quote from Drucker is from W. E. Deming, an engineer and statistician:

> It is wrong to suppose that if you can't measure it, you can't manage it – a costly myth.
>
> *(Deming, 1993, p. 35)*

We surely do not need to have an accurate measure of poverty before we can do something about it, but there is a germ of truth in here – although perhaps not in the way the apparent Drucker quote meant it. We can recognise poverty when we see it in front of us but few, if any, can manage to experience every community in the world. I have just made a quick count of the countries I have visited since birth and it comes to 30 (not including layover stops). Much of this has been work-related rather than holiday travel, and I must admit that my personal carbon footprint has not been helped by this travel, especially as I have visited some of the countries on the list many times. Although 30 countries might seem like a lot, it is not actually all that many when you consider that there are around 193 countries in the world at the time I am writing this, and most of them are in the first Olympic medal table I presented in this chapter (Table 1.1). Hence, I have been to just over 15% of the total; nowhere near even a quarter of the total. How can I hope to get a sense of important issues such as poverty on a global scale when I have not witnessed it for myself? I can watch the selective images and listen to the commentary of those places presented to me by others – the TV programme producers, writers and presenters – but, as exciting and informative as these may be, it is still their imagery and words that I am meant to use to formulate a vision of what can be a very distant place – distant in all senses of the word. I know that many readers of this book will be far better travelled than I am, indeed some may even make a hobby out of increasing the number of countries they visit, but also many will have travelled a lot less. It is hard to get an average number for every inhabitant of the world, but it is likely to be very low – probably only just above one country per person. Herein rests not just a conundrum but a problem. We live in an increasingly globalised world where both the decisions taken and the events that occur in distant places can have an impact on us all. We need to have a sense of what is happening 'out there', not just because it is interesting but because it matters to them and also to us.

This is where indicators come in. This book will present various pictures of the world using indicators, and these indicators are often presented to us by their creators in the same format as the Olympic medal table: As rankings of countries based upon these indicators. They are created by a wide range of different groups and all of them have two major goals in doing what they do. First, they want to tell us what is out there or, more accurately, what they think we should know about what is out there; and second, and more important from their perspective, they want us to do something about it. At its heart this is no different to the Olympic medal table that tells us the current state of play in terms of a country's success (or lack of it) in helping to get its athletes to the level where they win medals, and the message underlying it all is to encourage all countries to do better – even the country that comes top of the pile wants to remain there. The old adage that knowing one's place is a prelude to improving that place holds as true here as it does in any other context. What works for the Olympic Games also works in other areas, or perhaps more accurately it is assumed to do so by the people that create the indicators and league tables. But all the complicated issues we talked about with the medal table applies to all these other tables as well, without exception – whether the tables are for ranking poverty, corruption or even happiness, the same vagaries apply. Changing the methodology, no matter how compelling the argument for doing so, can change rankings significantly.

Therefore, the reader is encouraged to keep all of this very much in mind when looking at the examples given in this book. Just because it is a table of human development rather than one of gold, silver and bronze medals, and just because it was created by the United Nations Development Programme (UNDP), a widely respected body that works very hard to help alleviate the problems of poverty that exist today in the world, it does not mean that the issues surrounding the assumptions being made can magically disappear – far from it.

The point of the exercise in this book is, first, to give the reader visions of the planet that reflect important issues. Hence, the book is a collection of maps that have been shaded to show these issues as if they were somehow visible to us from a spacecraft orbiting the planet. The shading in such maps is another way of presenting tables: The depth of the shading equates to a value of the indicator and hence a place in the table. Figure 1.1, for example, is a shaded map showing the number of medals won in the 2012 Olympic Games as set out in Table 1.1. The darker the shading for a country then the greater the number of medals that it won.

The dominance of the northern hemisphere, most notably the USA, Russia and China along with Europe stands out in this map, as indeed is the almost absence of any shading at all for the countries of the African continent, parts of Asia and Latin America. Australia is a notable exception in the southern hemisphere.

Maps also help make comparisons much easier. Figure 1.2 is a map of the number of Olympic medals won per athlete (the values in Table 1.4). Again, darker shading represents a higher value.

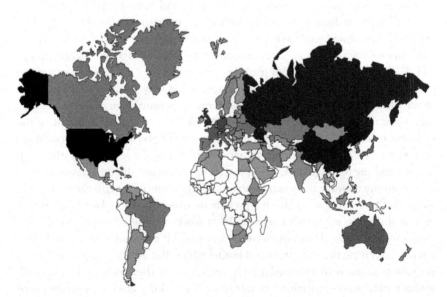

FIGURE 1.1 Total number of medals won in the 2012 Olympic Games in London. Higher values are represented by a darker shading

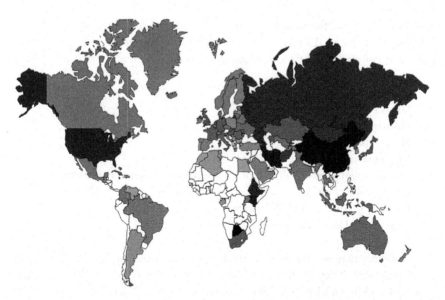

FIGURE 1.2 Total number of medals won in the 2012 Olympic Games in London per athlete entered by the country. In effect, this is the total number of medals won by a country divided by the team's size. Higher values are represented by darker shading

Flicking between the two maps gives a much better sense of the difference between these two measures of success, something that is much harder if looking at the two tables. What stands out here is the presence of some darker shading for parts of the world that were pale in the previous map. Most notable of all, look at Botswana just north of South Africa. That country came top in this ranking and its colour is the darkest shade of all those in the map. Therefore, from these two example maps I hope the reader will agree with me that maps are much easier on the eye than the tabulated listings shown earlier in the chapter. A picture really does paint a thousand words, and a map paints an indicator table with its multitude of numbers very well.

Second, I would like the reader to get a 'warts and all' understanding of what has gone into those pictures. After all, it is not just in this book that the reader will come across such things: They are quite literally everywhere. Magazines, newspapers, TV, the internet – all of these media use maps shaded in some way (or tables) to bring out differences. Indeed, it is hard to avoid them. It is all too easy to accept such devices without question, just as we can do with the Olympic medal table, without appreciating the assumptions that have gone into them – and these may be legion and not necessarily obvious.

I have chosen a number of indices, where an index comprises two or more indicators, as the foci in this book. They are not the only ones I could have selected, and neither are they the 'best' – whatever that may mean. I have chosen them because they span a number of important and in some cases interesting topics, and they also raise different issues in terms of their underlying assumptions. Just as with the Olympic medal table, I have no personal axe to grind in showing that a particular country does better or worse than others. The filters – indices – I have chosen for seeing the world are:

- Gross Domestic Product
- Human Development Index
- Happy Planet Index
- Corruption Perception Index
- Environmental Sustainability Index and Environmental Performance Index
- Ecological Footprint
- Poverty, Inequality and Vulnerability indices
- Sustainable Development Goal Index

These indices may not mean all that much to you now, but by the end of the book they will. They all have a technical feel to them, although some of the words are oddly familiar – happiness, corruption and vulnerability. Others like Gross Domestic Product and Footprint indices are obtuse to say the least while Corruption Perception Index, Environmental Performance Index and Sustainable Development Goal Index have at least some clues in their titles as to what they are intended to represent. But my intention is not simply to set out the various indices and what they are 'about', and neither is it just to show how the world looks through the lens of each of them. These days there are various sources available via the internet that allow the user easy access to that information and perspective. But I want to go deeper and look behind the indices and what they tell us of the world. Each of the innocuous looking names in the list above hides a long and, in some cases, tortured and heated history of debate, disagreement, consensus, rejection, acceptance and even anger. Each of them is built on a set of assumptions made by their creators for various reasons, and these can and have been highly contested. Also, there are the various efforts made over the years to relate the indices to each other and look at what that tells us about the world. But choices over which indices to relate and, critically, the conclusions that can be drawn from are also the source of much debate. For example, can we really say that the state of happiness in a country is primarily driven by economic performance? Those who think that we indicator geeks, or indeed scientists, are an unemotive lot would be astounded by the debates I have witnessed surrounding some of these indices: Television soap operas would not do them justice. I will tell the story of each of these indices, and it is a story full of more surprises and lessons.

Each chapter will cover one of the indices, and the later chapters in the book will explore how we have often tried to relate them each other and the pitfalls, and benefits, of doing so. Finally, I will set out my views as to where I think the science (and art) of indicators and indices could, and arguably should, go from here. Indeed, it is to highlight my view that they are as much art as science that I have given the chapters a musical theme, both with the headings and with the introduction.

Conclusion

Indicators are used by us every day, although for the most part they comprise visual clues rather than the 'hard' technical and numerical 'things' that we often associate with the label. But despite their objective and scientific appearance indicators do have a great deal of embedded subjectivity. This book is about that subjectivity – the soul of indicators and indices. It dwells less on the technical construction of the indicators and much more upon their history, what they were designed to do, the assumptions behind them and what they tell us of the world in which we live.

Note

1 The Olympic Games medal data were obtained from: https://www.olympic.org/olympic-results.

References

Deming, W E (1993). *The New Economics for Industry, Government, and Education.* MIT Press, Boston, Ma.

Magnay, J (2012). Team GB medal target for London 2012 Olympics is fourth place with 48 medals across 12 sports. *The Telegraph.* https://www.telegraph.co.uk/sport/olympics/9374912/Team-GB-medal-target-for-London-2012-Olympics-is-fourth-place-with-48-medals-across-12-sports.html.

Further reading

Girginov, V (ed.) (2013a). *Handbook of the London 2012 Olympic and Paralympic Games: Volume One: Making the Games.* Routledge, London and New York.

Girginov, V (ed.) (2013b). *Handbook of the London 2012 Olympic and Paralympic Games: Volume Two: Celebrating the Games.* Routledge, London and New York.

The following website sources are highly recommended for the reader who wants to go beyond the indicators I have included in this book:

The World Bank Open Data repository: https://data.worldbank.org/.
Our World in Data: https://ourworldindata.org/.

2

ECONOMIC INDICES

Introduction

It is probably true that there are more indicators that revolve around money than any other area of human endeavour. I say 'probably' because one only has to look at the news each day to read about economic growth (or not, as the case may be), how businesses are doing, household income, house prices, inflation and so on. Almost all these stories include accompanying indicators by which the journalists and those being interviewed seek to tell us how well, or badly, matters are progressing. It can be bewildering to see how all of them use these measures to make their case and dismiss the case made by their opponents, who in turn seem to use much the same indicators to spin a completely different picture. The head can swim as you hear that the country is doing very well – 'just look at indicator 1, indicator 2, etc.' – only to be told by the next person who appears in front of us that the 'country is in the doldrums as shown by indicator 3, indicator 4, etc.' what makes it worse is that the same 'thing' seems to have a number of indicators. In the UK, for example, we have two 'headline' indices of inflation:

- Consumer Price Index (CPI for short)
- Retail Price Index (RPI for short)

Both of them measure the change in prices of a 'basket' of goods and services over a period of time, and the index is expressed as a percentage: either a percentage increase or decrease. If inflation is positive (i.e. greater than zero) then the purchase power of money declines over time, meaning a pound buys less and less as its value declines. The CPI and RPI are, in essence, the averages of the change in price of the items in the basket, although part of this may well include an element of weighting to allow for the fact that a household may routinely purchase more of some goods (foods such as bread, for example) and services (such as water and power) than others (such as electronic goods). But why are there two measures? In fact, there are many measures of inflation, and, critically and obviously, much can depend upon what is included in the basket. As inflation gives us an idea how the purchase power of money changes over time, these measures are typically used by groups (pension companies, governments, service providers, trade unions, etc.) to estimate what they should be paid by customers, taxpayers, etc. for what they provide and also what payments (e.g. social security and pensions) should be made to households in order to maintain purchase power. But why have both the CPI and RPI? I am running the risk of oversimplification here given that there are variants of each, but the main difference between the CPI and RPI is that the CPI does not include the cost of accommodation (rent and mortgage payments, estate agent fees, etc.) while RPI does. Hence in the UK the CPI is often but not always lower than the RPI, and, perhaps unsurprisingly, those bodies that pay money to households (such as pensions and social security payments) tend to use the CPI, while bodies that require payment from households to government and private sector service providers often use the RPI. I suppose few have claimed that the world is fair!

The key point with the inflation story above is that indicator creators and users will have their own agendas, often political, and thus opt for those indicators that best reflect and support their position. It is not the indicators that are biased, they are just numbers arrived at via a formula, but those who create and use them who may have an agenda.

In this chapter we will explore some of the indicators often associated with 'money' or, more accurately, economics. We will begin the story with a depression.

The Great Depression and the birth of economic indices

The 1930s was a difficult decade, to put it mildly. It saw the 'Great Depression' from 1929 to 1939, with arguably the worst year being 1933. The Great Depression began with the stock market crash in 1929 in the US. The reasons for this were rooted within a period of rapid economic growth in the 1920s combined with a growing level of recklessness with regard to investments. People thought that investments were a guaranteed way of making money and the stock market would only go in one direction – up. Vast sums were being lent to companies and individuals without much thought about how they would pay the money back. But the economic fundamentals in the late 1920s were not as good as the surge in the stock market suggested and panic began to spread as that realisation slowly dawned. As we have seen with the more recent crash in 2008, which had many echoes in terms of causes to the crash of 1929, investors lost a lot of money, investment fell, companies (including banks) collapsed, governments had to step in to keep the banks going, unemployment rose and consumers stopped consuming. The result was a vicious cycle of decline that was, at the time, much worse in the 1930s than the earlier depression of 1873–1896. Although the 1929 crash started in the US it spread rapidly to other countries, especially those in Europe where the symptoms were the same. Indeed, no continent was untouched. A social security safety net provided by the state, as we see in many countries today, which helped prevent the worst effects of the 2008 crash, was virtually unknown in the 1930s and the result was a decline in living standards and a surge in poverty. The political responses across the globe were much the same as we have seen with the more recent economic collapse in 2008: A rise of populism and nationalism with charismatic and vociferous leaders taking power, often through democratic means, claiming to be able to fix all the ills at a stroke, often placing the blame on 'foreigners' and immigrants. For those who feel that they have been hurt by changes taking place outside their control in distant places of political power it can indeed be highly persuasive to hear that there are simple solutions just around the corner. The 1930s saw the coming to power of the Nazi Party in Germany and the awful ramifications that followed in terms of war and genocide.

It is not necessary here to go into the details of the Great Depression – the interested reader is referred to Robert McElvaine's (2009) authoritative and highly readable book on the subject – but there are points that need to be made in the context of indicators. First, politicians tended to think that the crash would run its course and all would be well, given enough time. Herbert Hoover was the US president between 1929 and 1933, at the height of the crash, and he believed that it would be short-lived, perhaps lasting just a few years. Hence his early interventions to deal with the problem were relatively mild, amounting to blaming Mexican immigration, a familiar refrain to this day, and asking leaders of industry to limit any lay-offs. A second issue was the reliance on the 'gold standard', which at the time fixed the amount of currency a country had in circulation. The US was a strong adherent to the gold standard in the years prior to the Great Depression and this prevented them from doing what many central banks did in response to the 2008 crash – print lots of money! Third, and probably hard to believe these days, there was a paucity of data collected by governments on how their economy was performing. I say it is hard to believe because we are all used to seeing economic statistics being reported in the media on a daily basis, but in the 1920s and 1930s it was different. This lack of appropriate indicators did not help politicians in framing their policy responses. The economist Richard Froyen makes the point:

> One reads with dismay of Presidents Hoover and then Roosevelt designing policies to combat the Great Depression of the 1930s on the basis of such sketchy data as stock price indices, freight car loadings, and incomplete indices of industrial production. The fact was that comprehensive measures of national income and output did not exist at the time. The Depression, and with it the growing role of government in the economy, emphasized

the need for such measures and led to the development of a comprehensive set of national income accounts.

(Froyen, 2009, p. 13)

The Great Depression did point to the need for a much more extensive system of national accounting to allow the better informed management of the economy. This is not quite the same sentiment as the supposed quote from Peter Drucker in the previous chapter:

> If you can't measure it, you can't improve it.

But it certainly points to the problems that can occur if you have little idea what is happening.

The Nobel Prize winner Simon Kuznets (1901 to 1985), a professor of economics at the Universities of Pennsylvania, Johns Hopkins and Harvard, is usually given the credit for the creation of our modern system of national accounting and indeed the science of economics. Among other things, he helped pioneer our modern concept of Gross Domestic Product (GDP), which forms the basis for this chapter. GDP is a term that is repeated so many times in the media that it is arguably the most reported and well known of all the indicators covered in this book, not least because politicians love to quote it when it suits them. Indeed:

> While the GDP and the rest of the national accounts may seem to be arcane concepts, they are truly among the great inventions of modern times. Much as a satellite in space can survey the weather across an entire continent, so can the GDP give an overall picture of the state of the economy.
>
> *(Samuelson and Nordhaus, 2010, p. 386)*

The comparison of GDP with the use of satellites is an intriguing one, and indeed we will return to this in the last chapter of this book. But it does have to be stressed that GDP, in itself, is not the answer to preventing economic crashes. If it were then the crash of 2008 would never have happened. Like all indicators it is nothing more than a tool – a window into the world of monetary flow – and what really matters is what people do with it together with the host of other economic indicators pioneered by Kuznets and his colleagues. Would the availability of the information captured by GDP have helped to prevent, or at least manage, the Great Depression and the decline of the world into war and genocide? After all, if the lack of such information was a contributory cause then would it have helped if it had been developed earlier? I cannot say I am convinced, although I accept that it may have helped. Given the lessons of 2008, where lack of access to all sorts of economic data, models and related indices can hardly be given as an excuse, it seems reasonable to assume that the answer is no. Few predicted the 2008 crash, least of all the economists and politicians, and they all had access to a wealth of data and models of the economy. No one can blame a lack of data or indices for the mistakes that happened. The crash of 2008 – dubbed by many the 'Great Recession' – was certainly the worst the world has experienced since the Great Depression and there are those who say it was worse than the Great Depression itself. Again, I do not wish to get into the details of a comparison between the two and the reader is referred to Barry Eichengreen's book, *Hall of Mirrors: The Great Depression, The Great Recession, and the Uses – and Misuses – of History*. Nonetheless, it has to be said that the Great Recession was arguably better managed than the Great Depression, and while we saw the usual blaming of immigrants, the rise of populism and nationalism, and the rise of civil and regional war, thankfully we have not as yet seen the rise of a new Adolf Hitler and moves towards a global war. What was different with the years after the 2008 crash was an injection of money into economies (the gold standard, which had prevented this for the Great Depression, had long been abandoned), the presence of social safety nets, immediate intervention to support the banking sector and, perhaps above all, the hard-won lessons learned following the rise of the Nazis and a desire amongst the vast majority not to return to those days. It would seem that indicators probably played a far greater role in the managing of the crisis and its aftermath than in helping to prevent it.

But how does GDP, praised by Samuelson and Nordhaus (2010) as being "among the great inventions of modern times", work? What does it measure? These questions will be addressed in the next section.

Measuring the size of an economy

Money, of course, flows between all of us in an economy. We all buy things, even if these days the exchange is electronic rather than by the use of physical cash. Similarly, we are paid for the jobs we do, even if it feels that we are not paid enough! Putting aside those value judgements, in basic terms we can think of an economy of which we are all a part as having two components – households and businesses – and four flows, as seen in Figure 2.1. In this diagram we can see that households buy things from businesses and in turn businesses hire people from the same households as labour. Hence there are two types of arrow in the diagram; one (represented by the thicker arrows in the centre) represents the flow of 'things' (goods, services and labour) while the thinner arrows around the outside represent the flow of money. Households receive money as salaries and wages while businesses receive money from the households when those households buy goods and services from them. Note that in this simple diagram we have no box for 'government'; we will return to that later. We also need to note that the economy in Figure 2.1 is 'closed', meaning that money does not flow outside the two groups – households and businesses. This is all very simplistic, of course, but even with this model it is apparent that we can measure the size of the economy in the diagram by either measuring the flow of money going to businesses (i.e. expenditures) or the flows to households (i.e. incomes). We do not need to measure both of these flows and add them together – that would

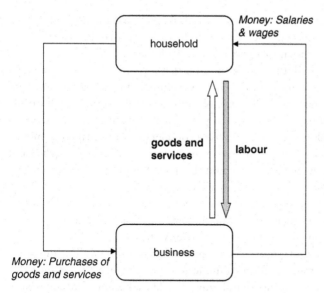

FIGURE 2.1   Simple model of an economy with only two components (households and businesses) and the flows of money, goods and services, and labour between them

be double-counting – as we assume that the money that flows into the businesses is the same money that flows out as salaries. Hence a measure of just one of the flows would be enough and economists speak of measuring flows using the 'expenditure' method (money paid to businesses) or the 'income' method (money received by households). There are various complicating factors to this model, as we shall soon see, but the fundamental idea of measuring the size of the economy in this way has much going for it.

But how do we measure expenditure and income in economies where there may be millions of households and hundreds of thousands (if not more) businesses? One option may be to use the information that households and businesses legally have to supply for the purposes of taxation – taxation on income, or taxation on goods and services sold. As noted above, the government is not included in the diagram, but it is obviously an important player in the economy as it is the government that receives the taxes and spends the money, we hope, on behalf of the wider society. Much depends, of course, on being able to receive the correct tax returns, and immediately you can see that there may be many flows that may be missed within contexts such as the 'black market' in goods, services and wages. After all, there are incentives for households and businesses not to pay taxes even if this is illegal! Going even further, there are other illegal monetary exchanges such as drugs and prostitution, to name but two, where the government will have little idea of the flows of money involved and, frankly, will legislate and enforce to stop them from happening. Thus, relying on data provided via taxation can only provide a partial picture of the economy, although estimates can be made of the size of the 'black market' and the illegal flows of money. Indeed, in some countries where the tax records may be sparse, there may be no option other than to 'guestimate' these flows.

The other problem with this simple model is that it is squarely rooted on flows of money for wages and consumption. There is nothing in the diagram that covers unpaid labour – voluntary work, for example, or indeed unpaid time spent caring for children and other dependants. Similarly, the quality of the goods and services provided to households is entirely absent from consideration. In a market economy it may be assumed the households are free to choose goods and services on the basis of quality, but all the diagram does is count the quantities of money flowing, and it makes no assessment at all about the quality of goods and services. But as long as we remember what is being measured here and do not try and pretend that it is anything more than that, then the assumptions are fine. The diagram is certainly not trying to capture any sense of quality of life or well-being.

While the simplicity of Figure 2.1 is appealing, in reality we do, unfortunately, have to make it a little more complicated. First, we have to include the government in this model, as shown in Figure 2.2. The addition of this one extra component certainly makes the model look far more complex.

Some of the complications of this new version come from the inclusion of flows of money to government via taxes – both on households and on businesses. In reality this can be highly complex as governments employ a variety of different types of taxes with all sorts of 'cut-offs' and incentives, but let us keep it

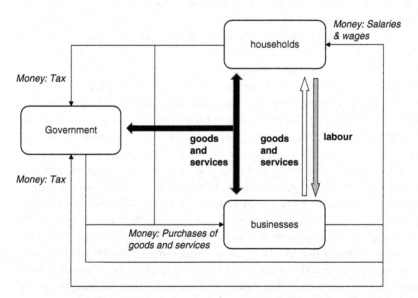

FIGURE 2.2 Monetary flow in an economy based on Figure 2.1 but including the government as a component. Note how the inclusion of this third component greatly increases the complexity of diagram

as simple as possible for now and just assume that money flows into the government's coffers. The government uses that money, of course, to purchase goods and services from businesses, to pay salaries to government employees, to give social security benefits to the poorest in society, and to make 'investments' in education, healthcare, infrastructure (roads, bridges, etc.) and so on. The black arrows in the centre of the diagram in Figure 2.2 represent these flows of goods and services from businesses to government but at the same time the provision of services etc. by government to households and businesses. Just as we did for the basic model with only households and businesses, we can assume that the income to government from tax is balanced by its expenditure on goods, services, salaries, investments, etc.

Even though Figure 2.2 may look confusing, there is yet another important element to consider – private investment. In reality, not all the money that households receive as income will be used for payment for goods and services and indeed payment to government as taxes. Many households and businesses will also save money in banks. The effect of this saving would be less expenditure on the part of households than may be expected based on their income. However, this is not how the real world works. You might intuitively think that this money is 'frozen' in the sense of being locked somewhere in a vault and therefore taken out of circulation. However, in reality, banks invest that saved money by lending it to businesses, governments and, indeed, to households, and they receive interest payments on those loans. Your savings are a source of income for the bank! The next version of the model (Figure 2.3) includes the banks, but I have had to simplify it. I have not included interest payments on loans (income for the bank) or indeed payments that banks make to their employees etc. However, I have included loans paid to businesses (not households) and savings made by households (not businesses). It could be assumed that the 'banks' component is part of

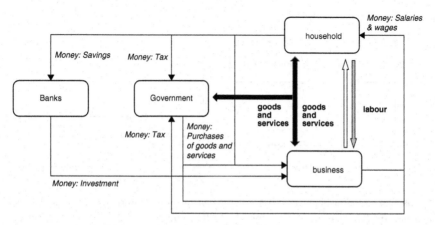

FIGURE 2.3 Cash flow in an economy. This diagram builds upon Figures 2.1 and 2.2 and includes banks and investment

the businesses box – as indeed many of them are – and it has only been separated out here to illustrate the flows of money as savings and investment.

Although Figure 2.3 is simplified, it still allows us to see that savings lodged with banks do not represent a removal of money from the system. Indeed, it can be reasonably assumed that banks have an incentive to invest as much of the savings as possible, while allowing, of course, for a constant stream of withdrawals from their customers. The banks need to invest that money carefully so it brings the best return for them, but that is another matter.

So how can we measure all these flows of money? If we focus only on expenditure and remove the physical flows of goods and services then the diagram can be greatly simplified once more, as shown in Figure 2.4. The thick black lines are the expenditures. C is the expenditure of households on goods and services, G is the expenditure by government for all of its provision (goods and services), including salaries of its employees, and I is the investment made by banks. It should be noted that I only covers investment made in the private sector, as investment paid for by government for education, training, infrastructure, etc. is part of G. The aim here, as before, is to make sure that flows are not double-counted as that would magnify the numbers involved.

We can assume that money flowing via the dotted lines in Figure 2.4 is in balance with the flows represented by C, I and G. Thus, households pay for goods and services, pay taxes and save money and these are covered by all of C, part of G and part of I. Businesses also pay taxes, which becomes part of G, and save money in banks, which becomes part of I. We can find C as businesses have to make tax returns on their sales, G is known from government accounts, and banks also return information on I as part of their accounts. Therefore, by adding C, I and G, and being careful about what we include within them, we can get an

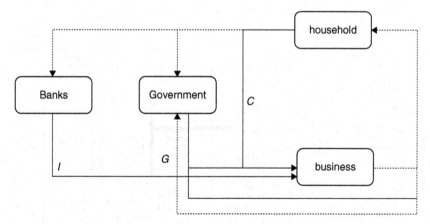

FIGURE 2.4 Components of GDP that cover private investment (I), government expenditure (G) and consumer expenditure (C). This is a stripped-down version of Figure 2.3, using I, G and C to highlight some of the key points that can be used to assess monetary flow. Note the focus on relatively few flows to avoid 'double-counting'

assessment of the total amount of money flowing in the economy – which tells us something about the size of the economy; albeit with all the caveats noted earlier.

The final point that needs to be made is that all these diagrams are for what we call 'closed systems'; no money can flow outside the boxes we have included. In reality, of course, this is not the case for most countries (if not all countries!). Businesses may sell their goods and services to households residing outside the diagram, in other countries for example. Similarly, households can also buy goods and services from outside the space represented by the diagrams. Thus, to get a better sense of money flow we have to adjust the calculation to allow for these flows of money into (via exports) and out of (via imports) the economy. The equation changes to:

$$\text{Size of the economy} = C + I + G + (EX - IM)$$

Where C, I and G are as above and EX = exports (goods and services are sold outside the economy and hence money flows in), and IM = imports (goods and services are purchased from outside the economy and hence money flows out).

It is important to note here that a good or service does not have to leave a country to be counted as an export. A good, and often quoted, example is provided by tourism within a country, which is actually treated as an export even though the goods and services are paid for locally, as tourists rent hotel rooms, hire taxis/cars, pay entrance fees to visit attractions, pay for meals, etc. This may seem counter-intuitive, as surely the money is being spent in the 'host' country. But while the payments may be local the money the tourists spend has come from outside the country; hence it is an export. Governments, of course, are often very keen to maximise the difference between exports and imports given that this has a positive effect on the size of the economy.

The equation and all the assumptions set out here are for what is commonly called the Gross Domestic Product or GDP for short. 'Gross' because it covers all the key flows of money in the economy, 'Domestic' because it is focussed on the components in the diagram that reside within an economy, and 'Product' because it is the sum of the five parts:

$$GDP = C + I + G + (EX - IM)$$

How does this play out in terms of real figures? Figure 2.5 is a pie chart of the GDP for the US in 2017, with the size of each segment representing the relevant proportion of the GDP for each component (actual figures are provided in billions of dollars). It may perhaps surprise you to see that the personal consumption component (C) comprised the bulk of the GDP for that year – at nearly 70% – while government spending (G) was only 17%. The (EX – IM) component was negative that year, representing a deficit in trade; that is, the US imported more than it exported, although the figure is not particularly large relative to the other components. It also needs stressing that the balance of GDP components you see in Figure 2.5 is not necessarily the same for all countries.

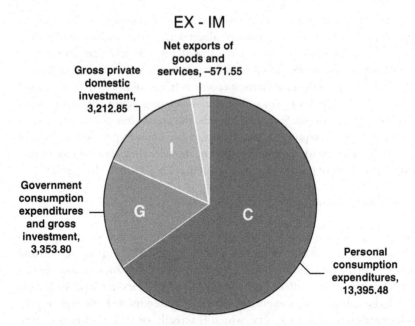

FIGURE 2.5 Split of the US GDP figures (US$) for 2017

Source: Own creation based on data from the World Development Indicators website http://datatopics.worldbank.org/world-development-indicators/.

Power to the people

As seen in the previous section, GDP is built upon many assumptions. But it is important to reiterate both what it is – a snapshot of the quantity of money in the economy based on the five components given above – and what it is not – it is not a measure of well-being or quality of life, and was never intended to be. We can make a leap of faith and assume that the higher the value of the GDP then the happier or better-off people will be, and we will come back to that point in Chapter 9, but this is a big assumption. Here we just need to reiterate that GDP is the measuring of money via C, I, G, EX and IM – nothing more and nothing less. That is all it was intended to be by its creators and that is what it is.

The various assumptions behind GDP were set out in the previous section, but one of them is perhaps not so obvious and, as a result, it is often overlooked. Much depends on the number of people in the economy as this will strongly influence the value of C as well as I (which matches savings) and G (which matches tax income). Businesses also pay tax and save money in banks, of course, so the number and size of them will be important, but, as we have seen with the US economy, the value of 'C' (consumer spending) can be a major factor and there is no getting away from the fact that GDP depends to some extent on population. Thus a country can have a very large GDP compared to another purely because it has a much larger population. This is not always the case, of course, as large populations comprising a high proportion of very poor people may not generate a large value for 'C', in which case that country would not have a large GDP, but clearly we nonetheless have an issue when trying to compare the GDP for the same country over time – has it grown simply because the population has grown? – or between countries that may have very different population sizes.

One simple way to accommodate the effect of population size is to divide the GDP by population to give the GDP per person or, as it is usually called, the GDP per capita; where 'per capita' literally means 'by heads'. As each living person has a head then per capita can reasonably be assumed to mean the same as per person! This adjustment depends, of course, on knowing the size of the population and while this may be known for many countries from regular censuses taken every 5 years or so, in others it is probably more of an estimate. Even if the population is not known to any great degree of accuracy an estimate based on projected growth rates will provide some degree of realistic adjustment. GDP per capita is widely used to compare economic performance over time for the same country and between countries. It is also used by some agencies to prioritise which countries should receive aid money.

Size matters; or does it?

If you think about it, what matters most of all is not the counting of money that you own but what you can buy with that money. Having a million notes of a currency might seem like a lot, but, if all it buys you is a cup of coffee, then it's not as good as it sounds.

Many of us know this only too well, as our salaries may increase at a rate far less than inflation. We may have the same money in our hands each month, but it seems to buy less and less. We call this 'purchasing power' and it will vary a lot depending upon where you are. Even in the place where you live, just think about how you can buy the same product or services for different prices depending upon where you shop. It can literally be the case that taking a few steps, from one shop to another one, can get you the very same product at a cheaper price. Thus we are all used to 'purchasing power' varying over space (e.g. between shops), but it also varies over time. A dollar now may not have the same purchasing power as it did some years ago, and indeed it may not necessarily be the same in the future.

Adjusting GDP in order to accommodate changes in the value of the currency over time is relatively straightforward; all we need do is allow for inflation or its opposite – deflation. If the inflation rate for a year is 5% then something that cost $1 at the start of the year will cost $1.05 at the end; meaning that the purchasing power of the dollar has declined. One approach is to standardise the value of the currency to that of prices from a chosen year, which can be thought of as a reference point. It does not really matter which year is chosen as all we are doing is recalculating the GDP for each year using prices from the reference year in order to track a trend in the size of the economy over time. Even so, international bodies recommend that the base year be changed every five years or so in order to ensure some consistency. The result is what economists call the 'real', 'constant' or 'chained' GDP with the year being used as the reference year clearly stated. This is differentiated from 'current' or 'nominal' GDP, which has not been adjusted to allow for inflation/deflation, and is based on market prices for goods, services, labour, etc. for that year – and of course this can change each year due to inflation.

A more challenging issue than dealing with inflation relates to comparing GDP across countries with different currencies and different social and economic contexts. This might, at first glance, seem like a trivial problem, as surely all we have to do is adjust the components of GDP in terms of exchange rate. It is common practice when going on holiday or a business trip to exchange money, even if it happens 'invisibly' when we pay for goods and services using a payment card. Currency exchange rates can fluctuate a great deal, of course, even within a single day, but it would be possible to generate an average exchange rate for the year and convert all currencies into, for example, US dollars. This is a logical step, but unfortunately it is only part of the picture. The problem is that a US dollar has very different 'values' in terms of what it can purchase across different countries. In some countries the local equivalent of a dollar can buy you a lot while in others it may not even buy you a cup of coffee. Therefore, a simple conversion of a country's GDP, be it 'constant' or 'current', into the equivalent of US dollars as a basis for country comparisons may be misleading, as it does not accommodate any differences in the purchasing power of the dollar. What is needed when comparing countries is not just a currency exchange but an adjustment for differences in purchasing power across countries; international agencies such as the Word Bank do just that and calculate 'purchasing power parity' (PPP) for all currencies used by nation states. While it sounds modern, the problem addressed by PPP has long been faced by people travelling between areas with different currencies and it has been going on for centuries. But how do agencies make the PPP adjustment? It boils down to finding the local prices of a 'basket' of goods that are as identical as possible across different countries. For example, one could use a loaf of bread made from wheat that is of the same size and composition and see what its price is in a number of countries. The chances are that even when converted to a common currency, such as the US dollar, the prices will be different and reflect differences in local cost of labour, raw materials, transport, machinery, etc. These differences, once allowance has been made for currency exchange, provide an estimate of PPP. The 'trick', of course, lies in the selection of goods that we use to make this comparison. In the case of a food such as bread, which is consumed in many countries and is usually made in that country rather than imported, any differences in price may reflect, at least in part, differences in the local cost of production. But this is not always the case, and countries where little wheat is grown may import the raw materials and so differences may thus also reflect an element of transport and storage costs as well as milling and production of the bread. Other goods, such as electronic goods or cars, may be entirely imported and differences in price may be a reflection of transport and retail costs as well as other factors such as demand. Thus the choice of an appropriate 'basket' of goods is critical and can heavily influence the values used to estimate PPP.

Once the value of PPP is known then currencies can be converted to an international standard. The one often used when comparing 'living standards' is the 'international dollar' or, as it is sometimes known, the Geary–Khamis dollar after its creators (Roy C. Geary and Salem Hanna Khamis). The international dollar, as the name implies, relates currencies to the US dollar for each year based on exchange rate and PPP. One of the very first published rankings of annual income within countries adjusted for purchasing power was produced by Colin Clark, a British/Australian economist, who was also one of the pioneers of national accounting. Figure 2.6 shows the world as represented by that very first set of income/capita data, adjusted for purchasing power, and it is, sadly, not too dissimilar from the maps the reader will see later in this chapter based on far more recent data. The largest incomes are in North America, Europe and Australia, although Argentina also has a relatively high income/capita. Africa and most of Asia, including China and India, have relatively low income/capita. Indeed, it is probably the rise of China and India as economies that generates the biggest differences when compared to modern data; in some ways the world has seen little change.

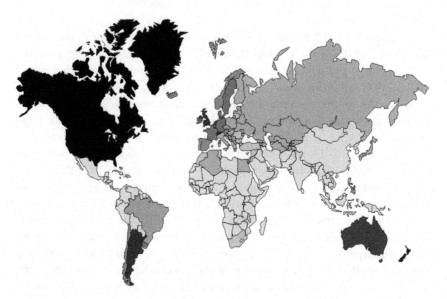

FIGURE 2.6   One of the first published rankings of countries based upon income/capita where income has been adjusted for purchasing power. The map is based on data provided by Clark (1940, p. 54) spanning the period 1925–1934

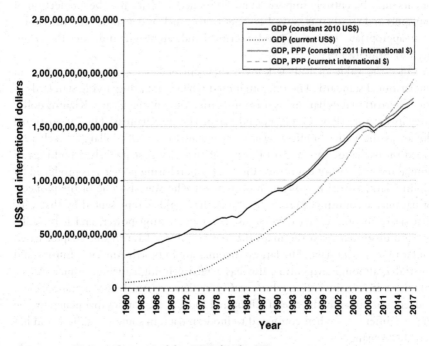

FIGURE 2.7 Various 'flavours' of the GDP for the US

Source: Own creation based on data from the World Development Indicators website
http://datatopics.worldbank.org/world-development-indicators/.

In Figures 2.7 and 2.8 you can see changes over time for some of the various 'flavours' of GDP for the US and Venezuela respectively. The pattern for the US is a gradual and smooth increase in all the GDPs over time with a minor 'blip' in 2008 as a result of the economic crash that year. Given the earlier discussion in this chapter, you may be a little surprised that the 'blip' is as small as it is, but even so this 'blip' did have serious ramifications for many people. Note how the figures for 'real' (or constant) GDP, where the dollar is adjusted to match its value in 2010, are higher than 'current' GDP up until 2008 when the lines cross. Some of the increase you can see in current GDP over this period is due to inflation. The GDP adjusted for purchasing power are identical to the unadjusted values for the simple reason that PPP uses the US dollar for the adjustment. The units for GDP adjusted for purchasing power are international US dollars. For Venezuela the GDP pattern is far more uneven, and perhaps even a bit chaotic, compared to the lines for the US, with various peaks and troughs that relate to numerous political and economic shocks experienced by the country, which continue to this day. For example, the dip in GDP around 2002/2003 was largely due to an all-out national strike, and the years that followed saw a number of currency devaluations. The increases you can see in GDP from the early 2000s were largely driven by a rapid increase in the global

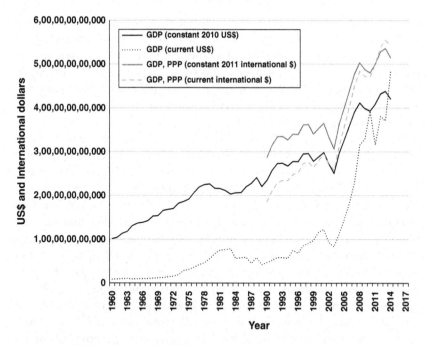

FIGURE 2.8   Various 'flavours' of GDP for Venezuela

Source: Own creation based on data from the World Development Indicators website http://datatopics.worldbank.org/world-development-indicators/.

price of oil. We tend to think of oil as coming primarily from the countries of the Middle East, such as Saudi-Arabia, but Venezuela is one of the world's largest oil exporters and has the world's largest reserves of oil and gas. From 2006 onwards oil accounted for some 90% of Venezuelan GDP. Also, in 2011, the government repatriated gold bullion previously held in London – although this was spent within a few years.

Battle of the acronyms: GDP vs GNI

GDP, in its various incarnations as noted above and including related measures such as GDP growth over time, is arguably the most commonly reported of economic indicators in the media but it is not the only one, and economists have come up with extensions of the idea of measuring money flows that take a wider perspective than the GDP focus on 'domestic' flows. For example, none of the GDP variants discussed above include other monetary flows into and out of a country, such as remittances. Households may well send money outside their country of residence as gifts, perhaps to friends and family members living in other countries, or, equally, they may receive such remittances. Multinational companies (companies with offices and factories in a number of countries), may also move money between their operations. Such movement of money across national borders is an illustration of the globalised world we now live in, but it is important to note that this is not the same as money being exchanged as payment for imports or as revenue from exports; there are no goods and services being exchanged and these figures are not part of the (EX − IM) component in GDP. The motivation for moving money around like this is varied, and as noted above for individuals it may often be to help friends and relatives. Companies, on the other hand, may move money for other reasons. For example, a company may move money into a country in order to be able to declare their profits there. This may seem an odd thing to want to do, but if a country has a low tax regime on company profits then there is an incentive for companies to report as much of their profit in that low-tax country as possible, even if only a small fraction of their sales take place there. Even relatively small differences of a few per cent in the tax rate can make a big difference if a company is declaring millions or perhaps billions of dollars of profit. This process of moving money around so as to declare it as 'profit' in low-tax countries may sound somewhat nefarious but may well be entirely legal, although it certainly raises moral questions. After all, other countries, including those where a large part of the sales have taken place, will lose tax revenue as a result of such 'movement', while the recipient countries will gain a lot for doing very little. You might not think that matters, but, as we have seen, governments invest in infrastructure, education and so on, which multinational companies benefit from and, if they avoid the payment of taxes, albeit quite legally, they are arguably benefitting from what the government has provided without making a fair contribution towards the costs. Once declared as profit for tax purposes the money may then be moved on again. We live in a world where moving vast quantities of money is very easy and quick to do.

Unsurprisingly, given the interconnected nature of the world economy, there has been an appetite to go beyond the 'domestic' focus of GDP and to try to capture the monetary flows (in and out) from around the world that are not part of (EX − IM). One such measure is called the Gross National Product (GNP) or Gross National Income (GNI) to distinguish it from GDP. GNP and GNI are similar in concept, although slightly different in the way they are calculated. The differences do not concern us here, but for both GNP and GNI the adjustment to the equation above for GDP can simply be set out as follows:

$$\text{GNI} = \text{C} + \text{I} + \text{G} + (\text{EX} - \text{IM}) + (\text{money flowing in} - \text{money flowing out})$$

The '(money flowing in − money flowing out)' component is called 'net factor income' and is in practice quite complex. If negative, then more money flows out of the country than into it.

While GNP and GNI (from here on I will just use GNI) arguably give us a better sense of monetary flow in an economy, given that all countries are intertwined in so many ways, it is important to note that the quantities of money flowing in and out of an economy may not necessarily be linked to the size of any of the other components in the equation. Thus, for example, while the GNI does not include flows of money as part of international aid, in some developing countries the flow of remittances coming into the country from citizens living and working overseas may be significant.

But how do the GDP and GNI values compare across countries? The answer to this question can be seen in Figure 2.9, which compares the two measures for 2017 adjusted to a per capita basis. The dotted line you can see represents equality between the two measures, with anything below the line meaning that GNI is lower than GDP while anything above it means that GNI is higher than GDP. Most countries appear to be on the line, which suggests that the two measures are broadly similar and the inclusion of 'net factor income' is not making a lot of difference, although the fluctuations either side of the line can be significant especially towards the left-hand side of the graph where the values for GDP and GNI are lower. Most of the countries at that end of the scale are less developed, and are mostly from Africa and parts of Asia. However, the major deviations from the line are to be found towards the right-hand side of the graph for countries with relatively high GDP/capita and GNI/capita. In the cases of Ireland and Luxembourg the lower GNI/capita compared to GDP/capita is caused by a net flow of money out of the country (net factor income is negative), caused in part by companies declaring their profits in those countries in order to pay less tax and then moving those profits out. This may seem counter-intuitive as

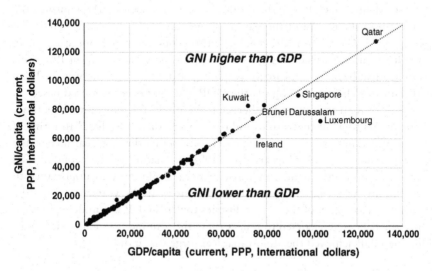

FIGURE 2.9   GNI compared with GDP. Dots are for individual countries and are the GNI/capita and GDP/capita for 2017. Figures are current international dollars based on adjustment via PPP

Source: Own creation based on data from the World Development Indicators website http://datatopics.worldbank.org/world-development-indicators/.

surely if money is recorded as flowing into the country and then flowing out again should the difference be relatively minor (just reflecting tax paid)? Well, not exactly. Some companies may have procedures in place that direct revenues to their subsidiaries in low-tax countries in such a way as to appear to be earned there, and this would contribute to the GDP of those countries. The result is an inflated GDP/capita figure for Ireland and Luxembourg, which might suggest that they are among the wealthiest countries in Europe if not the world, as you can see in the graph. But as money is being repatriated away from those countries once declared as profit, then it appears within the GNI with the result that the GNI/capita is lower than the GDP/capita. It all sounds incredibly convoluted, but the motivation to do this is strong given the relatively low corporation tax regimes (including various exemptions and inducements that go to make up the tax 'package') in those countries. Naturally all this has repercussions in terms of reputation, with some calling Ireland and Luxembourg 'tax havens', but also in terms of payments to bodies such as the European Union (EU), where the size of the payments is often based on GDP.

While the GDP and the GNI have their own logic, care does have to be taken to avoid double-counting. We have already noted this for the GDP in the diagrams above, and there are real dangers that the size of an economy may be overestimated. Internationally agreed guidelines exist for the calculation of both GDP and GNI and countries are expected to follow them. Changes in methodology do occur and these can result in significant changes to the recorded GDP and GNI. In 2014, for example, Nigeria changed its methodology for calculating GDP and the result was an 89% increase, which pushed the country into the top 24 of the world's league table of countries by GDP and the largest in Africa! How did it achieve this economic 'magic' in just a single year? It was done by changing the base year for constant GDP from 1990 (this reference had been used for many years) to 2010, but not in the way outlined earlier in this chapter. In Nigeria the GDP had long been measured based on estimates of industrial output rather than the use of the expenditure method. The contributions of the various industries were then 'weighted' in terms of their assumed importance to the economy. Changing the base year had a large effect because over the 30 years since 1990 the nature of the economy had changed significantly as new industries emerged and grew in importance to the economy while others diminished. Therefore, in the years after 1990 the contribution of many of the new industries towards GDP was underestimated. Once the data were changed to reflect these shifts in industry and weightings, then the GDP achieved the almost magical surge that it did.

The GDP picture of the world

As the reader will have seen from the previous sections, there are many ways of representing GDP. We can use current or real GDP, adjusted for PPP as well as adjusted on a per capita basis. The six maps in Figure 2.10 present the world as

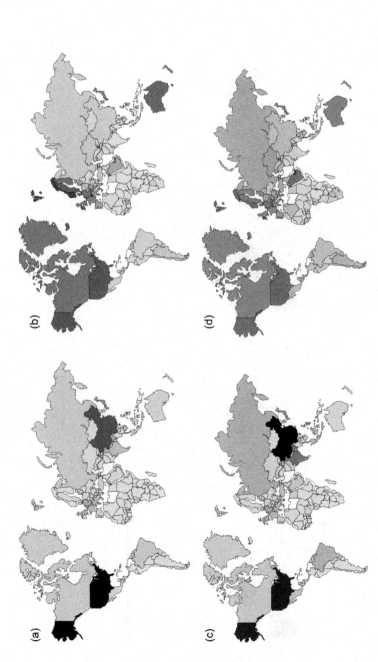

FIGURE 2.10 The world as represented by six indicators using Gross Domestic Product (GDP, 2016). (a) Current US$, (b) current US$/capita), (c) current international dollars (adjusted via PPP), (d) current international dollars/capita (adjusted via PPP), (e) constant (real) GDP US$ chained to 2010 prices, (f) constant (real) GDP international dollars chained to 2011 prices and adjusted for PPP

Source: Own creation based on data from the World Development Indicators website http://datatopics.worldbank.org/world-development-indicators/.

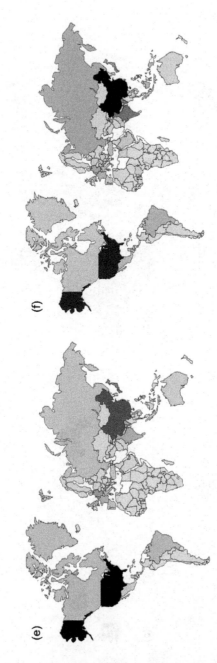

FIGURE 2.10 *(Continued)*

seen through some of this breadth of GDP representation, with all data taken from 2016. The following are some points for the reader to note.

Current versus constant (or real) GDP: The pictures of the world are much the same whether we use current or real GDP. The countries that dominate the scene are the US and China – the current economic powerhouses of the globe.

GDP unadjusted for PPP versus GDP adjusted for PPP: Again, there are no major differences in the world picture. The US and China continue to stand out.

GDP versus GDP/capita: This is where the global picture changes dramatically. Once GDP is divided by population size to give us GDP/capita then the world looks like a very different place. North America, Europe, Australia and Saudi Arabia now stand out as having the largest economies, while China and India are much diminished compared with their standing using only GDP.

Which of these pictures of the world is the right or the best one to use? There is no single answer to this question, and much depends on what you are trying to convey and why – and that is testament to the flexibility of the GDP indicator 'family' as well as its weakness. On the one hand we have the various adjustments designed to provide these different perspectives on measuring the size of economies, and all these adjustments have a logic that emerges out of the assumptions at play with the calculation of the GDP. But, on the other hand, this multiplicity of variants on the GDP theme allows anyone to select the one that they feel best suits the message they wish to convey. An interesting anecdote regarding this latter point stems from the recent (2016) Brexit referendum in the UK. It was a common claim during that referendum that the UK was the 'fifth largest economy in the world', with the insinuation that being such a large economy meant that the UK would be able to negotiate good terms when breaking away from the EU, and would also be able to flourish in terms of international trade. The assumption, as is so often the case with GDP, seemed to be that 'bigger is better'. But was it a true claim? Well, it depends on the choice of GDP, as shown in Figure 2.11. In this graph you can see the rank of the UK using four measures of GDP – current and constant (based on 2010) along with the PPP adjusted equivalents in international dollars. Also shown in the graph (the grey bars) are the numbers of newspaper articles that used the phrase "fifth largest economy" in relation to the UK. You can see that the use of this phrase surged in 2016, which was the year of the referendum. With the two PPP adjusted versions of the GDP, the UK was nowhere near the rank of fifth and was indeed only just in the top ten. It is only with the current GDP that the UK can be said to be ranked the 'fifth largest economy' from 2014 onwards, but that measure of GDP includes changes due to inflation. With the GDP adjusted to 2010 prices, so that inflation is removed, the UK was the 'sixth largest economy'. So the truth of the statement does indeed vary depending on which version of the GDP one wishes to use, and, of course, it suited those promoting Brexit to use the version that placed the UK in the most favourable position. Politics and economics are indeed intertwined.

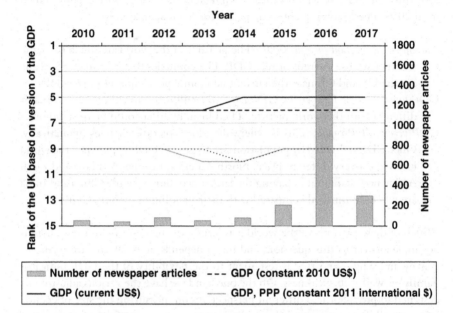

FIGURE 2.11 Rank of the UK in global league tables using four 'flavours' of GDP. Also shown in the graph are the number of newspaper articles that mention the term 'fifth largest economy in the world' in relation to the UK

Source: Own creation based on data from the World Development Indicators website http://datatopics.worldbank.org/world-development-indicators/. Newspaper article data from the Nexis database (https://www.nexis.com/).

Product or progress?

To end this chapter on economic indicators, I would like to bring you back to the points I made at the very start regarding the assumptions behind estimating the GDP. I noted the importance of good quality data from tax receipts etc. as the basis for estimating the money circulating in the system, but I also noted that much of what we do, and indeed value, is not 'receipted' in ways that mean it would be captured within the GDP. An example is voluntary work: Work that is provided without any payment exchanging hands. Voluntary work to help the most vulnerable in society or perhaps to help clean up the environment are all 'good' things and desirable. It is something that governments and indeed many of us like to see. But no matter how much voluntary work exists in an economy it is not captured within GDP, and neither should it be as there is no money being exchanged. This is not a criticism of GDP.

The limitations of GDP in the sense of it being a narrow indicator of monetary flow have been noted many times. Some have argued for GDP to be adjusted in ways that accommodate the various 'desirables' we need to encourage in society and, at the same time, highlight the need to remove 'undesirables', such as the cost of crime. I am not going to go into all these 'beyond GDP' attempts to design new indicators to use alongside GDP, but here I will present just one – the Genuine Progress Indicator or GPI, produced by an organisation called 'Redefining Progress', an environmental economics non-governmental organisation and 'think tank' based in San Francisco, US.

An example of the corrections that take place with GPI compared to GDP are shown in Figure 2.12 for the US (in this case for the year 2011). There are two pie charts in this figure, with the first of these showing the additions to the GDP – the 'desired' positives that GDP does not capture but that the creators of GPI thought should be included. You can see that the three largest additions are:

1. Benefits of housework
2. Benefits of consumer durables
3. Benefits of high education

Neither housework nor volunteer work are included within GDP as no money changes hands for these activities, let alone is recorded. But GPI assumes that they are valuable for society and so should be included. The inclusion of the other two components is perhaps more surprising, given that the purchase of consumer goods is included as part of 'C', and higher education usually results in higher salaries and hence greater expenditure. Why should they be emphasised in GPI? In the case of 'consumer durables', the key word here is 'durable' – meaning that they will last rather than having to be replaced on a regular basis. Perversely, the GDP is boosted by goods that are not durable and that have to be purchased regularly, as this adds to the flow of money in the economy. Hence GPI seeks to acknowledge the benefits to society of consumers buying goods that last longer rather than having to be disposed of every few years and dumped in landfill sites. The inclusion of the higher education component is intended to emphasise the benefits of education that go far beyond a simple ability for individuals to earn and spend more money. Higher education is regarded here as a wider good for society and not just a benefit for the individual; a point that, on a personal level, I would like to see more widely acknowledged and embraced by governments.

In terms of the negatives that GPI introduces compared to GDP, there are three that stand out in Figure 2.12.

1. Cost of inequality
2. Cost of non-renewable energy resource depletion
3. Cost of motor vehicle crashes

In the case of inequality, GDP per se is immune to this – it simply does not matter when calculating the indicator. A country can have a very high GDP if it has relatively few people with extremely high incomes and expenditures and everyone else is living at basic subsistence levels. Those who are 'super rich' will be earning and spending a lot of money, thus creating a high GDP. This may be

(a) Components which add to the value of GDP

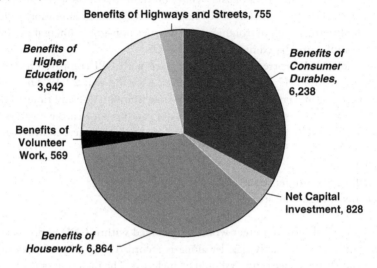

Benefits of Highways and Streets, 755

Benefits of
Higher
Education,
3,942

Benefits of
Consumer
Durables,
6,238

Benefits of
Volunteer
Work, 569

Net Capital
Investment, 828

Benefits of
Housework, 6,864

(b) Components which reduce the value of GDP

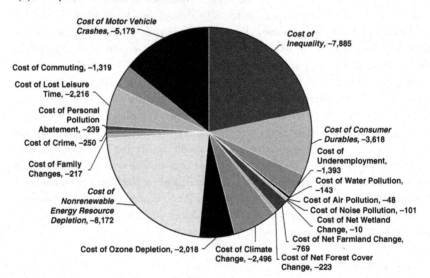

Cost of Motor Vehicle
Crashes, –5,179

Cost of
Inequality, –7,885

Cost of Commuting, –1,319

Cost of Lost Leisure
Time, –2,216

Cost of Personal
Pollution
Abatement, –239

Cost of Consumer
Durables, –3,618

Cost of
Underemployment,
–1,393

Cost of Crime, –250

Cost of Water Pollution,
–143

Cost of Family
Changes, –217

Cost of Air Pollution, –48

Cost of Noise Pollution, –101

Cost of Net Wetland
Change, –10

Cost of
Nonrenewable
Energy Resource
Depletion, –8,172

Cost of Net Farmland Change,
–769

Cost of Ozone Depletion, –2,018

Cost of Climate
Change, –2,496

Cost of Net Forest Cover
Change, –223

FIGURE 2.12 Corrections to the GDP (per capita) to generate the Genuine Progress Index (GPI) for the US in 2011. Figures are in US$/capita and come from Erickson and Fox (2018). (a) Components that add to the value of GDP, (b) components that reduce the value of GDP

Source: Own creation using data from Erickson and Fox, 2018. "Data for: Genuine Economic Progress in the United States: a Fifty State Study and Comparative Assessment", *Mendeley Data*, v1 (http://dx.doi.org/10.17632/c3xndx53td.1).

morally undesirable and hardly the stuff of a stable and vibrant society, but the calculation of a GDP is immune to such things. However, in a GPI the 'benefits' and moral desirability of a more equal society are acknowledged and rewarded by penalising its opposite – inequality. Items 2 and 3 are actually positives in a GDP – believe it or not! A country that depletes its resources may cause a great deal of environmental damage, including to others who live outside the country's borders, but this will be a positive in its GDP as it increases the flow of money. Similarly, car crashes are a positive in a GDP as they force people to spend money on repairs and new vehicles. But car crashes are obviously harmful in terms of the stress they cause, among other things. Therefore, both of these are seen as negatives for society and are reflected this way in a GPI. Indeed, a glance through many of the other smaller negatives in Figure 2.12 also raises a number of peculiarities. Crime, for example, is broadly a positive for GDP as people replace the goods that are stolen and spend money on insurance and crime prevention. The illegal drug trade and prostitution are also positives in GDP as they too increase the flow of money and, while they are not receipted directly, at some stage the money will find its way to places where it is receipted.

What difference is there between GPI and GDP? Time-series data to allow such a comparison are scarce for many countries but do exist for some, mostly in the developed world, and broadly show that GPI tends to be lower than GDP. Thus, when all the positive and negative adjustments are made along the lines set out above, the overall result is that countries become less 'wealthy' as measured by GDP. An example is provided in Figure 2.13 for the US state of Maryland, one of the first US states to embrace GPI as a key indicator of economic performance, with a strong interest from the state governor being instrumental in

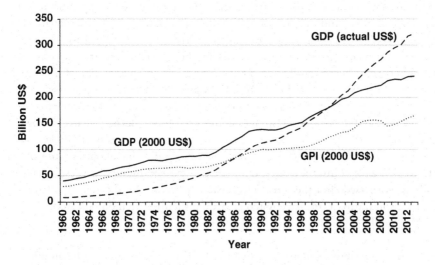

FIGURE 2.13   The GDP and GPI for the state of Maryland in the US

Source: Own creation using data from http://dnr.maryland.gov/mdgpi/Pages/default.aspx.

driving that popularity (Hayden and Wilson, 2018). There are three lines in the graph that span the period 1960 to 2013. One (the dashed line) shows the current GDP for the state and has been included here for reference given, that it is the one that often attracts the headlines in the media. The solid line is the constant GDP based on prices in the year 2000, and the The GPI, also based on 2000 prices, is shown as the dotted line. The GPI is always lower than its GDP equivalent based on 2000 prices, and indeed the gap between the two has steadily widened. One other interesting feature is the marked dip in the GPI around the 2008 financial crisis, which is not seen so clearly in the GDP (current or constant).

The problem with GPI is that it is more of a 'values-based' approach to measuring the economy and, of course, much depends on the values one wishes to promote. I have noted above that I agree with seeing higher education as being of wider benefit for society rather than just being an economic benefit for the graduate, but I acknowledge that not everyone sees it this way. For some, it would seem that higher education is nothing more than a commodity, and its value should only be seen in terms of economic gain for the graduate; nothing more, nothing less. They would probably argue against the adjustment for higher education being made in the GPI. If we change what should be promoted then the value of the GPI changes. Similarly, the assumptions behind various valuations, such as those of environmental pollution, ozone depletion, land-use change, etc. can be, and has been, challenged. If GDP can be thought of as more 'art' than 'science' then surely the same can be applied to GPI.

Another criticism relates to the use of GPI to influence policy. We have already seen how GDP emerged as a tool out of the maelstrom of the Great Depression, and many countries are now geared towards collecting the required data to calculate the GDP, even if some, as with the Nigeria example given earlier, use estimates of industrial 'output' rather than the more accurate expenditure (or income) method. Indeed, national accounts often report GDP on a quarterly (3 monthly) basis and there are moves to do this on a monthly basis in some countries. Much depends on data availability, but, increasingly, transactions are being recorded electronically and the data can be easily shared. The idea here is that regular publication of the GDP allows policymakers and managers to use the indicator to plan interventions. But such regularity is not yet possible for the GPI, although with enough resources committed to its routine calculation then this could be the case. This reflects to a large extent the origins of GPI within the research rather than the policy community. GDP came about because of a strong demand within the policy and management community as a result of the failings witnessed after the Great Depression, whereas GPI did not. As Hayden and Wilson (2018, p. 14) have noted regarding their research on GPI in the state of Maryland:

> An interviewee emphasized that the GPI 'wasn't designed to be an effective policy tool'; it was not developed in conjunction with policymakers or with them in mind.

This sense of indicators matching, or not in the case of GPI, the needs of those intended to use them is important and will emerge at various times throughout this book. It is an area that I feel has not attracted the attention it should.

Finally, and from a more philosophical standpoint, the monetisation of the environment can be questioned, as surely the value of the environment to us is more than something that can be reduced to money. Yes, people do pay to travel to scenic spots and pay a premium to live in places that they see as attractive, but that does not mean we can use such behaviour to allocate a monetary value to a nice view or a clean environment. Thus, some would argue that while GPI may be a well-meaning attempt to talk the language that policymakers often respond to, it is fundamentally wrong. What is needed, they would argue, is something far more radical, but that is another story.

Conclusion

In this chapter we have explored the origin and theory of GDP and how it has diversified into a variety of forms designed to address various limitations. But at its heart the GDP and its family of related indices do nothing more than assess the size of the economy. That is all is was designed to do and that is what it does. Even the adjusted forms of the GDP, designed to accommodate population size, inflation, purchasing power and monetary flows between countries, still do not change the fundamental nature of the indicator. Admittedly, the world can look quite different if one uses the various forms of GDP as a 'filter' but the beating heart remains the same – the indicator is what it is. There is nothing 'wrong' with the GDP, or any of its variants, as a tool, in the same sense that there is nothing 'wrong' with a hammer that causes nails to bend when hit. It is not the tools that can be a problem but the uses to which they are put by people.

The limitations of GDP have long been recognised and a 'beyond GDP' school of thought has existed for some years. One of the outcomes of this has been the attempt to generate modified forms of GDP, such as GPI, that take into account presumed benefits and negatives for society. However, GPI also has its limitations, in part born out of its creation by researchers to address what they thought was wrong with GDP.

References

Erickson, J and Fox M-J (2018). Data for: Genuine economic progress in the United States: A fifty state study and comparative assessment. *Mendeley Data*, Version 1. http://dx.doi.org/10.17632/c3xndx53td.1

Froyen, R (2009). *Macroeconomics: Theories and Policies*. Prentice Hall, Englewood Cliffs, New Jersey.

Hayden, A and Wilson, J (2018). Taking the first steps beyond GDP: Maryland's experience in measuring 'genuine progress'. *Sustainability* 10, 462. doi:10.3390/su10020462

Samuelson, P A and Nordhaus, W D (2010). *Economics*. 19th edition. McGraw-Hill, New York.

Further reading

Coyle, D (2015). *GDP: A Brief but Affectionate History.* Princeton University Press, Princeton and Oxford.

Eichengreen, B (2015). *Hall of Mirrors: The Great Depression, The Great Recession, and the Uses – and Misuses – of History.* Oxford University Press, Oxford.

Karabell, Z (2014). *The Leading Indicators: A Short History of the Numbers that Rule Our World.* Simon and Schuster, New York.

McElvaine, R S (2009). *The Great Depression.* Three Rivers Press, New York.

Raworth, K (2017). *Doughnut Economics. Seven Ways to Think Like a 21st-Century Economist.* Random House Business Books, London.

3

HUMAN DEVELOPMENT INDEX

Introduction

Despite the emphasis of the previous chapter, it is often said that life is not just about money. This is a refrain often heard from those of us who do not have much money, but one that we all content ourselves with as being true. There are other ways of looking at the world than through the monochromatic lens of economic measures like the GDP we encountered in the previous chapter. This is, of course, stating the obvious, but it perhaps says something about the world and its politics that economic growth has become the standard for so many to follow, and GDP growth is often at the heart of what many governments try to achieve and, indeed, how they are often judged by their electorate. There is a logic here in the sense that money can, of course, be used to purchase goods and services that can help to improve our lives. After all, healthcare, schools, police forces, roads, food, clothing and so on all cost money, as indeed do the luxuries, so it is foolish to claim that money is not important. We know, of course, that much can depend upon how money is distributed within a society and what the money is used for, but the logic that we must have the money in the first place in order to be able to afford good social policies, facilitate better distribution of wealth, etc. is compelling. Nonetheless, few, if any, would argue that GDP in and of itself is the only indicator a society should use to assess its progress. Economic indicators such as GDP may be widespread and form a strong focus for government action, but there are many other indicators that are also used, and we have seen one in the previous chapter – the Genuine Progress Indicator or GPI, which seeks to shift the emphasis to encourage 'desirables' and discourage 'undesirables' in the economy that are often not costed as part of GDP.

Development, or perhaps the lack of development, is one of those things that we all feel we know when we see it. We can all recognise the signals of poverty, for example, and signs that something is not right in society, such as civil unrest. This is a point that will be returned to later in the book, but for now, all we need say is that over the years there have been various efforts to come up with measures of how well a society is doing. In this chapter, we will explore one of them – the Human Development Index (HDI) – that was also designed to encourage a wider appreciation of the things that matter to us reaching out beyond monetary flows.[1] While not as old as GDP, HDI was first introduced in 1990, and thus has nearly 30 years of pedigree behind it and, perhaps unsurprisingly given the ambition of its creators, it has raised many issues and taught us many lessons.

When is development not development?

The term 'human development' has a very strange feel to it. At first glance it seems to equate to our 'growing up'; a development from baby to child to teenager to adult. Indeed, mention human development to many people and experience tells us that this sense of individual development from baby to child to adolescent to adult is what they are most likely to think. It takes some time to explain that 'human development' can have a different meaning from our biological and psychological development as individuals. In this chapter human development will be defined in terms of the words of those who helped champion the concept the most – the United Nations Development Programme (UNDP); one of the major United Nations agencies. They define human development in terms of what it is meant to achieve:

> Human development is a process of enlarging people's choices. In principle, these choices can be infinite and change over time. But at all levels of development, the three essential ones are for people to lead a long and healthy life, to acquire knowledge and to have access to resources needed for a decent standard of living. If these essential choices are not available, many other opportunities remain inaccessible.
>
> *(UNDP HDR, 1990, p. 10)*[2]

Or – perhaps to put this more succinctly:

> The end of development must be human well-being.
>
> *(UNDP HDR, 1990, p. 10)*

There you have it – human development is a process to achieve human well-being through enlarging choice. The HDR listed as the sources for the above two quotations refers to the Human Development Reports published annually by the UNDP since 1990.[3]

But explaining human development in the way that it is articulated above brings with it many other issues. To begin with, it seems oddly superfluous; like saying 'we want wet water'. Surely human societies have been developing ever since there were human societies and surely humans have always wanted to increase their well-being? From the days when humans began to create tools and migrate around the planet, they have adapted to their local environments and sought ways to improve their chances of survival and their levels of well-being. So why do we have to have a major United Nations agency tell us in 1990, and every year since, that this is important, and that to achieve it we must have a good education and a long and healthy life, among other things? Is that not obvious?

The term human development was coined at least in part out of a desire to 'push back' at what was perceived by some working in the development arena to be a dominance of economic-based thinking in development, as set out in Chapter 2. It was thus an attempt to reboot our relationship with monetary indicators. We have seen how GDP and its kin were developed as tools to help assess monetary flows in an economy following the Great Recession of the 1930s. However, a mindset has arisen that suggests money is the basis for all development; that nothing can be possible unless a country first has the money to achieve its development ambitions. Hence it seems reasonable to suggest that economic development is the bedrock of progress, and indicators such as those within the GDP family are key measures of progress (or lack of it, as the case may be). Towards the end of the Second World War, in 1944, when it was plain that the world would need a major phase of reconstruction following the horrors of that conflict, a meeting was held between officials of 44 allied nations at the Mount Washington Hotel, situated in an area called Bretton Woods, near the town of Carroll, New Hampshire, US, which led to the creation of the International Monetary Fund (IMF) and the World Bank. These 'Bretton Woods' institutions had clear mandates to help promote economic development, and the reports from these agencies (and others) subsequently became replete with long and detailed tables of economic performance of nation states. Whether this is fair, or not, the perception of some was that development was being defined and framed solely in terms of economics. Put simply, the aim of economic development should be to get countries with a low GDP to have a higher GDP.

During the 1970s and 1980s, many began to argue that, while an economic dimension to development is important, it should not be the be-all and end-all; it should not be the only way of seeing development. In fairness, it has to be noted that agencies such as the World Bank were not oblivious to these wider concerns and the need to think outside the 'economic box' when it came to development. In the mid-1970s, the World Bank began to concentrate its attention on supporting people in the developing world, and, in line with that change of emphasis, it began reporting a series of development indicators in its first World Development Report (WDR) of 1978.[4] In that report there are two tables that the authors call 'social indicators'; they "provide some information on health conditions and on the availability of health and education services". The 'health-related indicators' in the WDR of 1978 spanned measures such as life expectancy, mortality rates and access to safe water, while the education indicators included enrolment in primary, secondary and higher education and adult literacy rate. Thus, it would be wrong to think that the World Bank, at least from the mid-1970s, was solely focussed on economic development and only employed monetary indicators. However, it is probably also fair to say that the data required to underpin social indicators were sparse relative to economic data and indeed the WDR of 1978 laments the paucity of good quality data on many important social factors and calls for the collection of data to:

> [H]elp define the shortfall in meeting the basic needs of the population, is an urgent matter
>
> *(p. 73)*

The World Bank went on to publish a series of Social Indicators of Development (SID) reports between 1987 and 1996 that gradually expanded the set of social indicators beyond those reported in WDR (1978). As of 1997, these social indicators were included alongside the economic ones within a new annual publication – the World Development Indicators (WDI) report.[5] The WDI of 1997 was released the year before the setting of six 'International Development Goals' by the World Bank, United Nations and Organisation for Economic Co-operation and Development (OECD) and an accompanying set of indicators specifically chosen to assess progress towards those goals.

In parallel with the trends seen in reporting by the World Bank, the UNDP created its vision of 'human development' as noted above and began publishing its own 'human development' reports from 1990. Human development was meant to be more than just a collection of social development indicators, as in the World Bank series of reports, but an approach that set out people's capabilities to lift themselves and keep themselves out of poverty. Central to these capabilities was having an adequate income, the need for good quality education and healthcare. In that sense the HDRs were a marked departure from the World Bank reports even though it had become widely accepted since the 1970s that development was not solely about money.

Capturing capability: The Human Development Index

The Human Development Index (HDI) was in many ways the most tangible product of the UNDP's Human Development Reports. The HDI was designed by the UNDP to provide a headline measure of human development, just as the GDP and its ilk were the headline measures of economic development. GDP was regarded at the time, and probably still is, as a highly successful index (a collection of three indicators into one) that captured economic development. But while the World Bank and others had been publishing a variety of indicators of social development since the 1970s there was no equivalent single index that captured them all. Instead, the World Bank reports provided table after table covering disparate social indicators of education, life expectancy, water provision, etc. The HDI was designed by the UNDP to try to capture human development into a single index; one that would help refocus attention away from GDP. The HDI would thus be a 'champion' index – a banner – for human development.

The HDI was first presented to the world as a league table of nation states ranked in terms of their HDI value in the Human Development Report of 1990. Values of the HDI and new league tables based on that index have been released almost every year since then, and individual countries have moved up and down the HDI league tables as a result. While the HDI has evolved over the nearly 30 years of its existence, its core assumptions have remained intact and based upon the capabilities idea noted above. These are that human development can be captured via three components:

1. Income
2. Education
3. Health

Income provides means by which people can 'develop', education provides the means by which people can have opportunities to earn that income, and health is an obvious 'must have', so people can make the best use of their education and continue to earn an income. You can think of education and healthcare as the essentials for supporting opportunity, while income is an outcome of that opportunity – access to better-paid jobs, for example, or, also, income can be a means of providing better opportunities by having the money to invest in other income-generating activities. A focus on these three components does miss other things of course. For example, there is nothing in here about crime, leisure and happiness. It is all very well having good health, a good education and a good income, but does all this necessarily make you happy? They also seem to be quite individualistic; what about our engagement with other people? You could, of course, assume that it is 'given' as we do have to work together, whether we like it or not, to achieve a good education, healthcare and income, but this highlights that our societies are much more than the sums of their parts. Furthermore, there is nothing in the HDI that captures the costs of achieving these three components. The provision of education and healthcare services in terms of infrastructure (schools, hospitals, clinics, ambulances, etc.) and specialised professions (teachers, doctors, nurses) does cost money. Some of the income can be recycled in terms of taxes to help pay for all this, and governments can borrow money of course, but is all this being generated at the expense of damage to the environment and indeed working conditions? After all, we can decimate our natural resources to derive income for the government and support education and health provision. Hence there is nothing in the HDI per se that captures a sense of 'sustainability'; an ability to be able to continue having these things into the future without destroying all our natural resources or damaging the environment.

The UNDP, in fairness, has never claimed that the HDI captures all aspects of human existence, and while many people have called for variations on the 'HDI-theme' to accommodate what they see as 'missing' from the HDI, these have been resisted by the UNDP on the grounds that the HDI needs to be kept as simple as possible, along with a need for consistency, to allow for comparisons of country performance over time. Hence, the custodians of the HDI have espoused simplicity and consistency over and above capturing all aspects of something as complex and indeed subjective as human development. One can perhaps hardly blame them in the circumstances, and clearly there are trade-offs in creating an index to rival the economic indicators while at the same time keeping it as simple and consistent as possible. But there are important repercussions that arise from this decision that are often forgotten yet resonate throughout the pages of this book:

1. The HDI is a reflection of the 'humanity' of its custodians; hence, it is a human construct founded on assumptions.
2. The HDI is a simplification of human development; it only covers a few aspects of what its creators consider to be important.

The first point makes it clear that the HDI is a reflection of the views, biases, weaknesses, ambitions, desires, wants, needs, etc. of those who created and continue to maintain it. The HDI is made by people for other people to use. While we may agree with the inclusion of the three components listed above, this is not a 'given' for everyone and opinions do differ. It is not possible to say who is right and who is wrong, and indeed there are no 'rights' and 'wrong's here. While the HDI has an appearance of being scientific by virtue of it being numerical and being based on mathematical calculations that its custodians always set out in detail within each Human Development Report, this may create a false sense of objectivity. The mathematics is an expression of all the value judgements under point (1) expressed in a different language; that is all. All those biases, viewpoints, assumptions, etc. do not disappear just because we put them into numbers and equations. Unfortunately, we often lose sight of this when confronted with numbers and technical-looking formulae. We can become blinded to the imperfections.

Second, we have the reiteration of points made earlier about the simplifying assumptions made in the HDI; it misses an awful lot of what is important in our lives. Indeed, the very idea that we can capture the richness of human well-being within a single number for each country does sound somewhat ridiculous, and even within a single country of millions of people, there will of course be much diversity. It does have to be stressed that this is by no means unique to the HDI. We have already seen how even a far more focussed indicator such as GDP is also a simplification, as it fails to capture all the monetary flows in an economy; and also how changing the ways in which these are estimated can radically change the value of a GDP. Indicators and indices are, by their very nature, simplifications and the HDI is no different, although it could be argued that the creators of the HDI were perhaps too ambitious in trying to capture human development in a single index of just three components.

Finding the HDI

The three components of the HDI are:

1. Income: GDP/capita adjusted for purchasing power parity (international dollars)
2. Education: Enrolment in education (years)
3. Health: Mean life expectancy (years)

All three of these should not be all that surprising to the reader given the material we have covered earlier in the chapter and the focus of human development on:

> [A] long and healthy life, to acquire knowledge and to have access to resources needed for a decent standard of living.
>
> *(UNDP HDR, 1990, p. 10)*

None of these three components is especially new, even at the time the HDI was created in the 1980s. As noted in Chapter 2, GDP/capita has often been used as a measure of average income, while enrolment in education and life expectancy were among the social indicators included in the World Development Report of 1978 produced by the World Bank and also featured in the Social Indicators of Development reports published by the World Bank from 1987. Countries keep national accounts and hence GDP data are readily available. Most countries keep census data, even if they are estimates, for its citizens and this includes records of births and deaths, the data needed to estimate life expectancy, as well as population size, and it does seem reasonable to assume that life expectancy is a reflection of health. Records are also kept of enrolment in primary and secondary schools as well as universities and colleges. Thus, the HDI has tapped into data that are readily available and updated on a regular basis. Such data are not available instantaneously, of course, and there is always a time-lag such that published data typically reflect the situation some years before. Collecting data takes time and someone also needs to check and process them. Even with the communication and computing facilities we have available today these are still time-consuming activities. These time lags are often forgotten when we peruse tables of the HDI and it is a mistake to think that the HDI league table published in the Human Development Report of 2018 is a reflection of the situation in 2018; it is more likely to be a reflection of the situation in 2016 and 2017. Indeed, in the HDR of 2018 the HDI figures are reported to be from 2017.

There have been various complications to the simple triumvirate of income-education-health that underpin the HDI. The education component, in particular, has comprised a number of subcomponents over the years of the index. For example, in the HDI of 2017 the education component has two subcomponents:

Mean years of schooling, and expected years of schooling. The mean years of schooling subcomponent is defined by the UNDP as:

> Average number of years of education received by people aged 25 and older.
>
> *(HDR, 2016, Statistical Annex Table)*

Note that the definition refers to 'education' while the name of the subcomponent refers to 'schooling'. The subcomponent is capped for the calculation of the HDI and no country is allowed to have a value higher than the cap. For example, with the HDI of 2017, published in the HDR of 2018, the cap was set at 15 years and the UNDP claimed that this is the "projected maximum of this indicator for 2025" (HDR, 2016 Technical Notes, p. 2).

The expected years of schooling is:

> Number of years of schooling that a child of school entrance age can expect to receive if prevailing patterns of age-specific enrolment rates persist throughout the child's life.
>
> *(HDR, 2016, Statistical Annex Table)*

'Expectation' here is more in terms of what 'children' are likely to receive rather than expectation expressed as a wish, and in the calculation of the HDI this value is also capped. For the HDI of 2017, for example, it was capped at 18 years meaning that no country could have a larger value than 18 years. The rationale for this value in 2017 is given by the UNDP as follows:

> The maximum for expected years of schooling, 18, is equivalent to achieving a master's degree in most countries.
>
> *(HDR, 2016, Technical Notes, p. 2)*

Therefore, the rationale for a cap at 18 years goes beyond the 'school years' and includes time spent in higher education. However, by that stage people are typically no longer 'children' by most definitions.

For both subcomponents of education the minimum number of years was assumed to be zero. This seems fair as no one in the world should have less than 0 years in education. Nonetheless, it would seem odd to have two subcomponents for education – one of which appears to assess the actual years of schooling while the other assesses the number of years of schooling someone is likely to receive. The HDI of 2017 uses an average of these two, presumably to arrive at a best estimate for each country:

$$Education\ component\ of\ the\ HDI = \frac{Mean\ years\ of\ schooling + Expected\ years\ of\ schooling}{2}$$

While the example given above is for the HDI of 2017, and the same rationale has been applied to other published values of the HDI since that year, there have been various other expressions of the 'education' component in the HDI since its first release in 1990. I will not go into all of them here, but all that needs to be reiterated is that the education component has changed over time, although, in fairness, the creators of the HDI have always provided a rationale for any changes they introduced.

Another complication with the HDI is provided by the income component. For many of the versions of the HDI produced each year since 1990 this has been reflected by GDP/capita adjusted for purchasing power parity (PPP), and we covered this measure in some depth in Chapter 2. The rationale behind the use of GDP/capita as an estimate of income may look rather simplistic, but it is a relatively easy value to obtain given that most countries routinely publish their GDP and undertake a population census every few years. The use of PPP to adjust the GDP also has a logic given that a dollar has very different 'purchasing power' across the globe; in some places it will not even buy you a cup of coffee while in others it can buy you a full meal. More recent versions of the HDI have tended to use Gross National Income (GNI) rather than GDP, and the difference between these two measures was also covered in Chapter 2. While the GNI and GDP may not be all that different for many countries, there are some, where the difference can be significant. It is worth bearing in mind that this shift from GDP/capita to GNI/capita for the HDI between 2009 and 2010 can have an impact on the final value of the index and potentially affect ranking.

While the shift from GDP to GNI is an example of a change in the calculation of income, the complications with how income is represented in the HDI do not stop there; and from the very origin of the index they appear to stem from a desire on behalf of the UNDP to avoid income 'swamping' the other two components. This is understandable given that the range of GDP (or GNI) per capita between countries is much higher than it is for life expectancy and education. Indeed, differences in education and life expectancy pale when compared to the differences seen in income, as we also saw in Chapter 2. For example, in the HDI of 2017 the maximum life expectancy was 84 years (Hong Kong) while the minimum was 52 years (Sierra Leone). The average life expectancy reported in that HDR was 72 years. The highest mean years of schooling was reported to be just over 14 years (Germany) while the lowest was between 1 and 2 years (Burkina Faso); and the average figure globally was 8.5 years. These are large differences, and they may shock you. The life expectancy for Hong Kong is 32 years higher than that for Sierra Leone, although the disparity is less than double. By way of contrast, the mean years of schooling for Germany is some 12 times greater than that for Burkina Faso and Niger. But in the same dataset used for the HDI, the highest income value is more than 176 times higher than the lowest; a difference that is far higher than we see for life expectancy and also much greater than we see for mean years of schooling.

From the very beginning, the creators of the HDI recognised that such a large disparity in income could have a distorting effect on the index and they introduced ways to limit this influence. One simple approach is to provide a ceiling for income. For the HDI 2017 the cap they imposed was $75,000 with a minimum of $100. Therefore, any country with a GNI/capita exceeding $75,000 was 'pulled back' to the cap and any country with a GNI/capita less than $100 was 'uplifted' to that value. The caps allied to the income data have changed over the years of the HDI although, in fairness, the UNDP always provide a rationale for their chosen caps.

However, capping the GDP and GNI at the top and bottom of the ranges was deemed by the UNDP not to be enough and indeed has its own disadvantage in the sense that much depends on where the caps are artificially set. After all, at one extreme we can use caps to force all the countries in the world to have exactly the same income! Hence, the UNDP have opted to further reduce the large variation in income between countries by 'compressing' the data. Unfortunately, the custodians of the HDI over the years have never quite made up their minds as to the way in which the income data should be 'squashed'. The most common approach they have applied is to take the logarithm. The reader is probably used to the idea of logarithms to the base 10. Here:

$10 = 10^1$ and logarithm is '1'
$100 = 10 \times 10 = 10^2$ and the logarithm is '2'
$1000 = 10 \times 10 \times 10 = 10^3$ and the logarithm is '3'
$10,000 = 10 \times 10 \times 10 \times 10 = 10^4$ and the logarithm is '4'
$100,000 = 10 \times 10 \times 10 \times 10 \times 10 = 10^5$ and the logarithm is '5'

And so on. In the list above, the values have gone from 10 to 100,000, a difference of 99,990, but the logarithm for these values has only changed from 1 to 5. For example, the logarithm of the maximum value ($75,000) allowed for by the UNDP in the HDI 2015 is 4.88, which you can probably guess from the sequence above, given that 75,000 is between 10,000 and 100,000 but nearer to the 100,000 end of that range. Hence, we would expect to see a logarithm under 5 but nearer to 5 than 4. But the logarithm for the minimum GNI/capita ($100) is 2. Therefore, higher values for income/capita will be 'pulled' down and the overall effect is to minimise the difference between the countries.

The use of logarithms is not the only approach applied by the UNDP to 'squash' income data after the caps have been applied, but over the 30 years' history of the HDI it has been the most common. For the HDIs that use the logarithm to 'squash' the range from low to high income/capita, the UNDP used what is called logarithm to the base 'e' rather than base 10 shown above. Logarithms to the base e, where 'e' is shorthand for Euler's number (approximately 2.71828), named after the Swiss mathematician Leonhard Euler, are referred to as 'natural' logarithms. The term 'natural' here means that the logarithm is based on a number that is mathematically meaningful rather than the use of an arbitrary number such as 10, and for a variety of reasons it is favoured by technicians and researchers, but the 'squashing' effect for higher values is the same as using logarithms to the base 10. For example, here are the 'natural' logarithms (or LN for short) of the numbers listed above:

10: LN = 2.3026
100: LN = 4.6052
1000: LN = 6.9078
10,000: LN = 9.2103
100,000: LN = 11.5129

The final step in the process of calculating the HDI involves an adjustment of the data per country, once caps and other adjustments have been applied, so that each of the three components has a value between 0 and 1. After all, the three components have very different units. GDP or GNI per capita is measured as international dollars, education and life expectancy as years. We can't lump together dollars and years, that would be like adding apples and oranges, but we have to adjust them in such a way that the units disappear, and they are all on the same scale. This is a common issue with indices, and there are various ways it can be addressed. The one used by the creators of the HDI is commonly used, and indeed will re-emerge with various indices covered in this book. It involves the notion of a 'distance to target'. In essence, we calculate the difference between the value of the component for a country and some notional 'target' of what we would ideally like the value to be; hence it is an aspiration. The target can be set by experts based on some rationale or it could be found by simply taking the maximum value for all the countries in that dataset. Thus:

$$Distance\ to\ target\ for\ a\ country = target\ value\ (maximum) - country\ value$$

The closer the value for the country is to the target then the smaller the 'distance to maximum'.

Alternatively, we can define the minimum value (or find it from the data) and look instead for a distance away from the minimum:

$$Distance\ to\ target\ for\ a\ country = country\ value - target\ value\ (minimum)$$

The larger the value then the further away a country is from the defined minimum and vice versa.

Either approach achieves the same goal, but the distance to target does need to be placed into the context of the range (maximum value − minimum value) for each of the three components of the HDI, given that, as noted above, their ranges are very different. If we have defined either the maximum or minimum values as targets, or maybe both, then those are the values that we use to define the range. Hence, we now have what can be called a 'standardised' value.

For example, if we defined the minimum and maximum values that a component should have then:

$$Standardised\ value = \frac{country\ value - defined\ minimum\ target}{\left(defined\ maximum - defined\ minimum\right)}$$

The numerator is the distance to target while the denominator is the range. In effect, we have rescaled the distance in terms of the largest distance possible in the data, and the values will vary from 0 to 1. If the standardised value is 0 then the country value matches the minimum, while if it is 1 then it is as far away from the minimum as it can be.

For the HDI, the UNDP have adopted various approaches to setting the maximum and minimum targets over the years; sometimes they set it themselves while on other occasions they have used the maximum and minimum values from the dataset. When the targets were set by the UNDP they have always provided a rationale for their decision. For example, for the life expectancy component in the HDI of 2017 the assumed maximum value was set at 85 years with an assumed minimum of 20 years: A range of 65 years.

> The justification for placing the natural zero for life expectancy at 20 years is based on historical evidence that no country in the 20th century had a life expectancy of less than 20 years.
>
> *(HDR, 2016, Technical Notes, p. 2)*

The standardised value for life expectancy was found by:

Standardised value for life expectancy

$$= \frac{country\ value - defined\ minimum}{\left(defined\ maimum - defined\ minimum\right)}$$

$$= \frac{country\ value - defined\ minimum}{85 - 20} = \frac{country\ value - 20}{65}$$

For Norway, the country with the highest HDI 2017 and a life expectancy of 82.3 years, this works out as:

$$Standardised\ value\ for\ life\ expectancy = \frac{82.3 - 20}{65} = 0.959$$

Thus, Norway is some 96% of the distance along the desired range for life expectancy.

As can be seen from the example above, the calculation of the three standardised values is relatively straightforward arithmetic, and easily achieved with a spreadsheet, with the key decisions being more in terms of selecting the maximum and minimum values. Change the assumptions and the values of the standardised value will also change. For example, rather than 85 and 20 years let us continue to use 85 as the maximum but this time use 50 as the minimum. I can justify my decision for using 50 years as this is actually only just below the minimum life expectancy in the dataset (52 years, for Sierra Leone), and therefore, I could argue, more 'realistic'. Now we have:

$$Standardised \ value \ for \ life \ expectancy = \frac{82.3 - 50}{85 - 50} = \frac{32.3}{35} = 0.923$$

Now it seems that Norway is 92% of the distance along the desired range rather than 96%; a small difference in achievement, just 4%, but a difference nonetheless. However, as we are using the same maximum and minimum values for doing the rescaling for all the countries then it will not affect their relative ranking in terms of life expectancy; each country ends up in the same position.

Finally, we take the average of the three standardised measures to give us our HDI, which will range from 0 (no human development) to 1 (maximum human development) with all the countries spread out between those extremes. The rule of thumb is that the higher the value of the HDI then the better the level of human development in that country; and vice versa. But while the arithmetic might look simple and beguiling, remember that what matters here is the setting of the targets, and that was not done by a machine but by people.

Please note that the three components of the HDI have consistently been weighted equally by its custodians. This assumption was hard-wired into the first version of the HDI, published in 1990, and has remained in place ever since. Indeed, it is the assumption of the equal weighting of the three components that created the need to adjust the income component so strongly. However, while equal weighting is the simplest course of action, one can certainly question this assumption. Does it seem fair to weight health the same as the other two, given how fundamental it is to our lives? After all, without good health it is hard to see how we can flourish so surely health must be weighted more strongly than the other two? But this is a very subjective 'call' and I have no doubt that others will disagree with that view.

It is also worth noting yet another apparent oddity at this point. For some of the HDIs, those published before 2010, the UNDP implemented the equal weighting by taking the arithmetical mean of the 'standardised' values of health, education and income:

$$HDI = \frac{Health + Education + Income}{3}$$

This is indeed what most people would regard as the 'mean' or 'average': We add up all the values and divide by the number of values we added together (in this case 3). It is the simplest and most intuitive approach to take. But there are other types of 'mean' besides the 'arithmetic' and since 2010 the UNDP has opted for what is called the 'geometric mean' of these values, where:

$$HDI = Health \times Education \times Income^{1/3}$$

Or

$$HDI = \sqrt[3]{Health \times Education \times Income}$$

As you can tell from the equations, there is a big difference between the means generated using the arithmetical or geometric approaches. For example, let us assume that for a single country each of the three HDI components has a value of 0.5 (after standardisation). This means that in each case the country is place halfway along the defined target range. Taking the arithmetic mean, we have:

$$HDI \left(based\ on\ arithmetical\ mean \right) = \frac{0.5 + 0.5 + 0.5}{3} = 0.5$$

As we would expect, the HDI turns out also to be 0.5. Taking the geometric mean:

$$HDI \left(based\ on\ geometric\ mean \right) = \sqrt[3]{0.5 \times 0.5 \times 0.5}$$

And:

$$HDI \left(based\ on\ geometric\ mean \right) = \sqrt[3]{0.125} = 0.5$$

Here the two methods for calculating a mean generate exactly the same result; all well and good. But this is a special case, and if we vary the values of the three components as shown in Table 3.1 then the outcomes differ significantly.

Note how, using the arithmetic mean, it is possible to substitute a poor performance in one component (in Table 3.1 it is life expectancy) with a better performance in another (in this case income). Despite the life expectancy component dropping from 0.5 to 0.1, which is, of course, a very bad change indeed, the overall HDI calculated using the arithmetic mean remains at 0.5, as the increase in income provides a compensation for the decline in life expectancy. Thus, the HDI based on the arithmetic average can hide compensations between the components, and somehow this does not seem to be right. The use of the geometric mean avoids this issue as it is sensitive to a decline in any of the three components, even if apparently compensated for by a corresponding increase in another. In the table we can see the HDI based on the geometric mean does decline as life expectancy goes down, even if there is an increase in income.

The 'soul' of the HDI has remained the same over its nearly 30 years' history in the sense that it has consistently been based on three components deemed

TABLE 3.1 Results obtained for the Human Development Index (HDI) using the arithmetic and geometric means of the three component values

HDI version	Standardised values of HDI components			Calculated HDI using different types of mean	
	Life expectancy	Education	Income	Arithmetic	Geometric
1	0.5	0.5	0.5	0.5	0.5
2	0.4	0.5	0.6	0.5	0.49
3	0.3	0.5	0.7	0.5	0.47
4	0.2	0.5	0.8	0.5	0.43
5	0.1	0.5	0.9	0.5	0.36

Note: There are five versions of the HDI in the table, differing in terms of the life expectancy and income components. The life expectancy component declines from version 1 to version 5 of the HDI while the income component increases. With the arithmetic mean the HDI remains the same (0.5) across all versions, as a decline in one (life expectancy) is compensated for by an increase in another (income). However, with the geometric mean the HDI declines from 0.5 (version 1) to 0.36 (version 5).

to be important for human development, and the UNDP has resisted repeated calls for the index to undergo a more radical change. However, this should not hide the fact that, underneath that surface of consistency, there has been a series of changes about the precise nature of each component, how the raw data are adjusted, and how they are put together (integrated) to generate the HDI that we see as the 'headline' table in the Human Development Reports, which is picked up and reported by the press and others. Countries will move up and down the HDI league table depending on how they perform in each of the three components, but such changes can also be the result of changes in calculation. Again, and in fairness, it has to be noted that the UNDP is not unaware of the changes that can occur to the HDI and country rank based upon shifts in methodology, and in more recent HDRs they have also published versions of the HDI in other tables based on a consistent methodology in addition to the headline HDI for that year. However, and unsurprisingly, it is the headline HDI that – well – tends to grab the headlines!

The world according to the HDI

While the reader would no doubt have picked up the message in the previous sections that the HDI is just one way of representing the world, and indeed is founded on many assumptions and choices that have changed over time, let us be neutral about the details and instead see what the world looks like using the index.

Figure 3.1 presents the world according to the HDI in 1990 and 2017; a gap of 27 years (more than a quarter of a century). As you may probably have expected, the picture presented of the world is a mixed one, with some countries coming

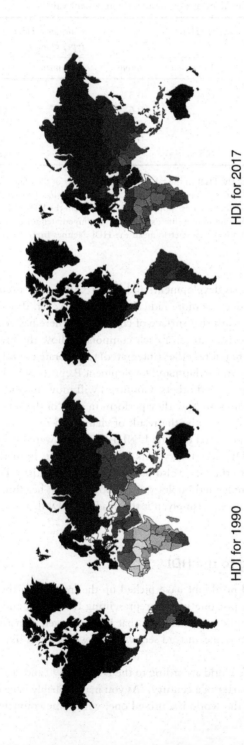

HDI for 1990

HDI for 2017

FIGURE 3.1 The world according to the Human Development Index (HDI) of 1990 and 2017. Darker shades equal higher values for the HDI, which equate to better human development

Source: Own creation based on data from the Human Development Reports (1990 and 2017).

out badly while others do well. Africa as a continent tends to do badly in terms of the HDI, and indeed in both 1990 and 2017 countries from that continent have tended to dominate the bottom half of the HDI league tables. Indeed, the three components of the HDI tell us a very consistent story.

Figure 3.2 shows the world as it appears through the three component 'lenses' and it is frankly hard to tell them apart. Admittedly the three maps you see here are based on the standardised values of the components, each spanning a range of zero to one, after the adjustments noted above had been made. Hence, we would expect to see a lessening of the variation we would see with the raw data, especially for the income component. For example, a visual comparison of the GNI/capita map with those of GDP/capita (adjusted for PPP) in Chapter 2 do present quite a different story. Admittedly, the maps in Chapter 2 are of GDP rather than GNI, and as already noted these are not the same thing, but more important is that the maps of Chapter 2 are based on the original data with no 'capping' or adjustment by taking logarithms. It seems that whatever the lens we use, the HDI story remains the same and, given the consistency of the story presented by the three HDI components, it is easy to see why the map of HDI 2017 looks the way that it does.

However, what is perhaps the remarkable, and indeed most disturbing, aspect of Figure 3.1 is the consistency between 1990 and 2017 rather than the consistency between the components of the HDI. The lighter shades (lower values of the HDI) are concentrated in the same places. Seen through the lens of the HDI the world of 2017 looks very much like the world of 1990. We can also check for consistency by plotting the values of the HDI 2015 against those for 1990, as shown in Figure 3.3, with each dot on the graph representing the HDI for a country, and the pattern is a clear one – the two are indeed related. It seems that a quarter of a century has not been anything like enough time to fundamentally disturb the human development order of the world. However, the graph does show up some good news that is perhaps not so obvious from the maps. Look how the pattern of dots has moved up. This suggests that while the order has broadly stayed intact, the HDI values for the countries in 2017 have increased compared to those of 1990, and this applies to countries at the low end of the scale as well as those at the upper end.

We can see this change between 1990 and 2017 more clearly by ranking the countries according to their value of the HDI, as shown in Figure 3.4. The difference between the top and the bottom countries in HDI 1990 was much greater than it was in HDI 2017. This suggests that, slowly but surely, the world is becoming a better 'HDI-defined' place. While the relative imbalance between the northern and southern hemispheres remains over those 27 years, the world has seen an improvement in human development as the lowest performers are gradually catching up with the best performers. It's just that those countries with lower levels of human development have not improved enough over those 27 years to bring them in line with those countries that have higher levels of human development – but there is progress nonetheless and that is encouraging.

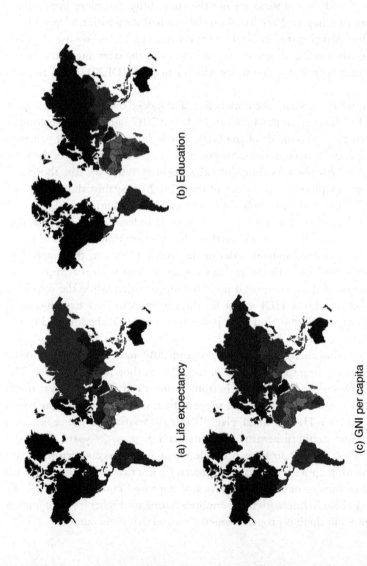

(a) Life expectancy

(b) Education

(c) GNI per capita

FIGURE 3.2 The world according to the three components of the Human Development Index for 2017. Darker shading equates to higher values

Source: Own creation based on data from the Human Development Report (2017).

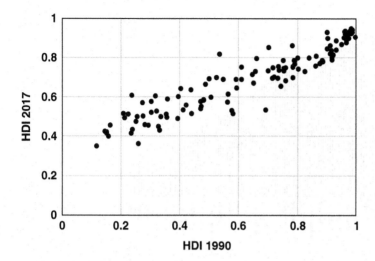

FIGURE 3.3 Relationship between the HDI 2017 and HDI 1990. Each dot in the graph represents values for a single country

Source: Own creation based on data from the Human Development Reports (1990 and 2017).

The human development of the world is improving but the differences between low and high have been incredibly resilient to change. We are not yet at the point where all countries are equal when seen through the eyes of the HDI, but we are getting there. As the title of this chapter implies, it may be a sad world after all given that many countries, especially in Africa, still have relatively low levels of the HDI, but perhaps we have good cause for optimism.

Conclusion

The history of the HDI provides many facets of interest to an indicator geek such as myself. These range from the motivation behind its creation as a 'counter indicator' to the perceived dominance of GDP, which was the flagship indicator for economic development along with its various incarnations that emerged over the years as the methodology and assumptions behind it changed. Perhaps of greatest interest, at least for me, is the use of an index to galvanise action by those with the power to bring about change. The HDI was designed to bring about change in the world and help improve the lot of millions; its design had purpose. The league-table style of presenting the HDI was also implemented on purpose and was part of that desire to bring about change; governments would feel a pressure to improve their standing. The HDI can certainly be criticised for its oversimplicity and omission of factors such as the environment; and the constant changes in methodology over the 27 years of the life of the HDI can be seen positively, as a flexibility to adapt, or more negatively as sowing confusion and

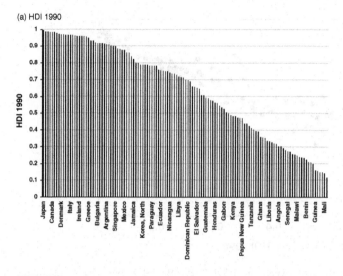

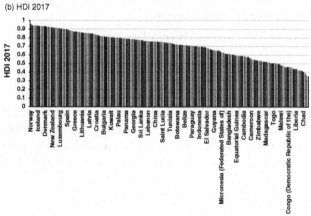

FIGURE 3.4 Distribution of the Human Development Index (HDI) for 1990 and 2017 across countries, with higher values of the index on the left-hand side of each graph and lowest values on the right-hand side. Note how the difference between high and low values of the HDI is much 'steeper' in 1990 compared with 2017

Source: Own creation based on data from the Human Development Reports (1990 and 2017).

diluting the message. But whatever the criticisms, the creators and proponents of the HDI have tried to make a difference for people. Whether the HDI has been instrumental in the change we have seen in the world measured in terms of the index (Figure 3.4) is another matter, and it can be challenging to separate out the impact of a single index from the many other factors at play.

Notes

1 You can find technical information on the HDI on the UNDP website, including technical notes and tabulated annexes. It is these technical documents and tables that have furnished a number of the quotations used in this chapter. http://www.hdr.undp.org/en.
2 UNDP Human Development Report for 1990 (and all subsequent ones) can be accessed at the following website: http://www.hdr.undp.org/en.
3 Most of the references in this chapter are to the UNDP Human Development Reports and the World Bank series of reports. The UNDP Human Development Reports can be accessed at the following website: Human Development Reports: http://www.hdr.undp.org/en.
4 The World Bank series of development reports can be accessed at: http://www.worldbank.org/en/publication/wdr/wdr-archive.
5 The World Development Indicators provided by the World Bank are accessible here: http://wdi.worldbank.org/tables.

Further reading

The Nobel Prize-winning economist, Professor Amartya Sen, was highly influential in the creation of human development as espoused by the UNDP. Two of his books in particular are well worth reading for those interested in the theoretical foundation of human development:

Sen, A (1999). *Commodities and Capabilities*. Oxford University Press, Oxford.
Sen, A (2001). *Development as Freedom*. Oxford University Press, Oxford.

For an historical analysis of human development and the HDI I can highly recommend the following:

Hirai, T (2017). *The Creation of the Human Development Approach*. Palgrave Macmillan, Cham, Switzerland.

A number of books have been written on the World Bank and the following text provides some interesting insights on its historical development since its birth in the mid-1940s:

Marshall, K (2008). *The World Bank. From Reconstruction to Development to Equity. Global Institutions Series*. Routledge, Abingdon, Oxfordshire.

4

ECOLOGICAL FOOTPRINT

Introduction

The notion of a 'footprint' representing an impact is probably as old as the human race. It has been said that one of the most popular poems ever written in English is Henry Wadsworth Longfellow's 'A Psalm of Life' and in it we find the words:

> Lives of great men all remind us
> We can make our lives sublime,
> And, departing, leave behind us
> Footprints on the sands of time.

Over thousands of years we have, as a species, certainly had a big impact on the planet upon which we live, and that impact continues to this day. We can actually see many of those impacts from orbit including the effects of infrastructure and agriculture, while others may be less obvious. Carbon dioxide and many other 'greenhouse' gases are colourless and can't be seen with the naked eye, and while some pollution of our water systems can be seen from space, there are others that cannot. At the time of writing this book there are increasing concerns about plastic pollution in the oceans, and, while it is possible to see the accumulations of such waste on the surface, this material does degrade over time; however the chemicals and particles that are released can be highly damaging to ocean life. All this represents humanity's visible and invisible footprint on the planet.

The use of the footprint analogy has found its way into the world of indicators, and we see it today in two forms – the 'Ecological Footprint' and the 'carbon footprint'. They sound similar, but are in fact quite different. The Ecological

Footprint is designed to encapsulate humanity's broad impact on the planet in terms of the resources that we use to live and to support our quality of life, and is expressed in terms of the surface area of the planet needed to support an individual based on what they produce and consume. By way of contrast, the carbon footprint is a measure of the release of greenhouse gases (e.g. carbon dioxide and methane) associated with a product or process. The carbon footprint is expressed in terms of carbon dioxide equivalents (or CO_2e), which reflects the fact that there are many gases that have the capacity to act as greenhouse gases, and, while carbon dioxide tends to receive all the bad publicity, there are others that may be present at much lower concentrations but be far more effective at keeping heat in. Methane, for example, is a greenhouse gas often released as a pollutant in industrial processes and agriculture, and it is far more effective at keeping heat in. The emission of 1 tonne of methane is equivalent in terms of global warming potential to 25 tonnes of carbon dioxide (Wright et al., 2011). Remarkably, there is one gas, sulphur hexafluoride, used for a variety of purposes, including as an insulator in the electrical industry, which is 22,800 times more potent than carbon dioxide as a greenhouse gas (Wright et al., 2011). Thus, 1 tonne of sulphur hexafluoride has the equivalent greenhouse effect of 22,800 tonnes of carbon dioxide.

The carbon footprint is important in terms of climate change and has received much political and media attention, although arguably nowhere near enough. Efforts are now being brought to bear on reducing our use of fossil fuels by substitution with other renewable energy sources such as solar and wind. We can also calculate the 'embodied' carbon in a manufactured product or service and that provides a basis for reduction, perhaps by introducing new processes. However, calculating the carbon footprint for a product or service can be complicated, especially as we must consider the carbon emissions from 'cradle-to- grave'. For example, when we consider the carbon footprint of a car it is not just its manufacture we have to consider but also its maintenance and disposal at the end of its useful life; every step in that cradle-to-grave sequence has a cost in terms of carbon emissions, and disposal at the end of the product's life is no different. However, calculating the embedded carbon footprint flowing from all these processes is necessary but challenging, and I do not wish to take the reader down that road (pardon the pun). The interested reader is referred to the excellent introduction to carbon footprinting by Mike Berners-Lee entitled *How Bad Are Bananas? The Carbon Footprint of Everything* (2011). For more technical treatments of the carbon footprint and detailed case studies, the reader is referred to The *Carbon Footprint Handbook* (Muthu, 2016) and *Carbon Footprint Analysis* (Franchetti and Apul, 2012).

In this chapter I will focus entirely on the Ecological Footprint, largely because it is much broader than the carbon footprint and raises a number of interesting points regarding its calculation. Having said that, the Ecological Footprint does include a 'carbon footprint', but it is considered in a different way to that outlined above.

The theory of the Ecological Footprint

The Ecological Footprint (EF), as a measure of our impact on the planet, was developed in the early 1990s by the academics William Rees and Mathis Wackernagel. Together they published a book in 1996 that set out the theory and practice of the EF (Rees and Wackernagel, 1996). Values of the EF are now published by a non-governmental think tank, based in California, called Global Footprint Network (GFN). One member of GFN is the World Wide Fund for Nature (WWF) and another is the New Economics Foundation (NEF), and these regularly publish values of the EF produced by GFN. The NEF uses the values of the EF in its calculation of the Happy Plant Index, which we encounter in Chapter 7. Indeed, the EF is often used in indicator research as the 'cost' to be compared to a benefit – such as economic performance (e.g. with GDP) or human development (e.g. with the HDI).

There are two ways of thinking about the impact (footprint) that we have on the planet. First, we can think in terms of the impact caused by the production of things that we need such as food; and, second, we can think of it terms of the impact caused by our consumption of things. On a global scale these are balanced, of course – by and large what we produce is what we consume – but there are differences across countries in the sense that in some places production may be greater than consumption, and indeed vice versa. Given that we produce and consume many things from animal- and plant-based products as well as using energy to do the growing and processing, it is perhaps not surprising that finding the production and consumption versions of the EF for any country is a complex process involving many assumptions. Here I will continue to follow the general principle set at the start of this book of avoiding the detailed mathematics involved, but as with all indices much does depend upon how it is done and the assumptions that are made. Therefore, while wishing to keep the technical details to a minimum, it is nonetheless necessary to set out the basic assumptions.

The first point to make is that the EF has six components based upon the demand for a fundamental resource – the surface area of the planet. The surface area includes land and water and is, of course, a finite resource. It is possible to build in three dimensions (up/down as well), and it is not unreasonable to think that, in the future, it may possible to make better use of our limited surface area, but even so the fundamental unit is still the land (or water) surface and this is fundamentally constrained. Our planet has the following areas (taken from Wikipedia):

Total surface area: 510,072,000 km²
Land area: 148,940,000 km² (29.2% of the total)
Water area: 361,132,000 km² (70.8% of the total).

(https://en.wikipedia.org/wiki/Earth)

Most of the planet is covered with water and, in the above list, 'land area' comprises a variety of different types, including forests, built-up areas, deserts and mountains.

Of the total of 148.94 million km², it was estimated (in 2015) that some 37% was used for agriculture (crops and grazing); a figure that is perhaps surprisingly high. Forests are estimated to cover about 31% of the total land area. Thus some 68% (two-thirds) of the land area of the planet can be considered to be 'biologically productive' meaning that it "supports significant photosynthetic activity and the accumulation of biomass" (Lin et al., 2017, p. 56). However, not all the forest area will be utilized by humans – although one of the most significant issues we have faced in recent years has been the destruction of natural forest and its replacement by agriculture and plantation crops such as oil palm. Such a change does bring with it a significant decline in biodiversity even if the land now produces more of the things (e.g. palm oil) that we want. Of the remaining land surface, the vast bulk of it is classified as desert but that includes all barren areas such as the polar-regions (Antarctic and Artic) along with the more familiar 'sandy' deserts like those of the Sahara and Arabia. By definition these desert regions have relatively low biologically productivity. Given all the megacities on the planet, it again might surprise the reader to know that urban land coverage is but a fraction of a percentage point. But, of course, all these percentages are in a state of flux. Our cities are growing but probably more important is that technology can allow for an expansion of crop and grazing areas into areas currently considered to be barren. Climate change will also have an impact. Nonetheless, the land surface area of 148.94 km² can be considered to be an upper limit of resource for producing biomass of use to us. Having said that, of course, we must not forget the potential of the seas. Most of the fish we catch (some 70%) currently comes from the Pacific Ocean with 20% coming from the Atlantic. The oceans do vary in terms of their 'biological productivity', and this is again in a state of flux due to climate change and pollution.

The EF focusses on the biologically productive area of the planet, both land and water, and conceptualises it in terms of six types of use:

1. Cropland
2. Forestland
3. Fishing grounds
4. Grazing land
5. Carbon uptake
6. Built-up land (urban areas and dams)

Numbers 1, 2, 3, 4 and 6 are self-explanatory. 'Carbon uptake' is the area (both land and water) required to absorb the greenhouse gas called carbon dioxide produced by humanity, primarily via industry, electricity generation (coal, oil) and transport. Note that 'carbon uptake' as expressed in the EF is not the same as the carbon footprint discussed briefly at the beginning of the chapter, although they are both addressing the same issue and one can be converted to the other. After all, we can think in terms of the carbon dioxide released into the atmosphere (carbon footprint measured using CO_2e) or what it takes to absorb that carbon dioxide from the atmosphere (carbon uptake).

The EF employs an assumed 'biological equivalence' for these different land-use categories and is a way of redefining the Earth's surface in terms of its productivity, that is defining it in terms of producing biomass for the benefit of humans; be it for food, forest products or absorption of carbon. For example, grazing land is considered to have a relatively low biological productivity as grass is converted to animal biomass, and this rate of conversion from the energy in sunlight captured by the grass to meat, milk and other animal products is not high. The rate of conversion of solar energy to biomass is typically of the order of a few per cent (1 to 2%) for crop plants, sugar cane has one of the highest conversion rates at 4%, and grazing animals would only convert a small percentage (<1%) of the solar energy captured by plants for their biomass (Gliessman, 2007). Putting these two together, the overall efficiency of conversion from solar energy to meat products may be a small fraction of 1% (Gliessman, 2007). The same rationale is applied to fisheries. On the other hand, cropland is considered to be more productive than grazing land and fisheries, as most of the products are directly used for human food. Thus, simply adding these areas together in terms of biological production would not work; it is like adding apples and oranges. In order to allow such addition these areas have first to be converted into a standard unit of area, and the EF creators have opted for something called 'global hectares' (gha). The conversion between actual areas under these land-/water-use types to areas based on productivity is achieved using a set of equivalence factors (EQVs), which differ between land-/water-use types but are assumed to be the same for all countries. Table 4.1 gives an example of this conversion between actual areas, using 2016 data from the United Nations Food and Agriculture Organisation (FAO), and their equivalent global hectares. The table shows a range of equivalence factors that reflects changes in how they were calculated based on the availability of new information and changes in assumptions. Taking the 2017 EQV values these can be multiplied by the 'use' areas in 2016 to give the equivalent biologically productive areas in global hectares. For two use types in particular, fisheries and grazing land, which have been shaded in Table 4.1, there is a notable reduction in the biologically productive area, largely reflecting the biological cost of converting primary energy (sunlight captured by plants and phytoplankton) to animal biomass. For crop- and forestland the equivalent biologically productive areas are larger than the actual areas, reflecting their ability to harness sunlight and convert it to biomass.

How are these equivalence factors in Table 4.1 chosen? After all, a great deal would seem to depend on the selection of these conversions and, as can be seen in Table 4.1, there has been some variation over time. The creators of the EF do provide a rationale for the choice of these values and the various assumptions made by them. For example:

> The equivalence factor for built up area is set equal to the equivalence factor for cropland, reflecting the assumption that built up areas occupy former cropland.
>
> *(Lin et al., 2017, p. 54)*

TABLE 4.1 Land-use areas (2016) and equivalence factors for the ecological footprint

Surface area use	Actual area (billion ha)	Equivalence factors (gha)				Equivalent area (billion gha) (using 2017 equivalence factors)	Difference in area (actual area − equivalent area)
	2016	2005	2006	2007	2017		
Fisheries (marine and inland)	2.89	0.40	0.41	0.37	0.35	1.01	**−1.88**
Land uses							
Crops	1.73	2.64	2.39	2.51	2.52	4.35	2.62
Forest	4.21	1.33	1.24	1.26	1.28	5.38	1.18
Grazing	3.67	0.50	0.51	0.46	0.43	1.58	**−2.09**
Built-up area/infrastructure	0.17	2.64	2.39	2.51	2.52	0.43	0.26
Carbon uptake		1.33	1.24	1.26	1.28		
Total	**12.66**					**12.75**	

Note: Equivalence factors convert actual areas of each land-use type to equivalent areas in terms of their biological productivity. For carbon uptake the assumption is to use the values for forest while built-up area/infrastructure uses the value for cropland.

Source: Equivalence factors are from GFN while the 2016 areas are from the United Nations FAO.

And

> The equivalence factor for marine area is calculated such that a single global hectare of pasture will produce an amount of calories of beef equal to the amount of calories of salmon that can be produced on a single global hectare of marine area. The equivalence factor for inland water is set equal to the equivalence factor for marine area.
>
> *(Lin et al., 2017, p. 55)*

Anyone can agree or disagree with these assumptions, of course, and indeed the methodology used by GFN to find the equivalence factors. After all, changes to the equivalence factors will alter the results of the EF. It can also be questioned why these equivalent factor assumptions apply to all places, and indeed there have been various efforts to calculate more location-specific values. For example, Liu et al. (2015) have published equivalence factors for China and their values were as follows:

Fisheries: 0.35
Cropland (and built-up land): 1.71
Grazing land: 0.44
Forestland (and carbon uptake): 1.41

In Table 4.1, the fishery and grazing values are similar to the 2017 global values, but the cropland value for China of 1.71 is significantly lower than the global equivalence factor of 2.52. While I accept that this may look somewhat confusing, and potentially open to bias if each country can set out its own equivalence factor and thereby reduce its apparent footprint on the planet, the key here, as with all indices, is transparency. As long as the methodology and any assumptions are set out clearly and in detail, then they are open for anyone to critique, evaluate and modify.

Setting out key assumptions, such as the values for the equivalence factors, is an important aspect of the EF. Once these have been made, then the calculation of the EF for each country is, in essence, a detailed accounting exercise that draws heavily from secondary datasets, such as those provided by the United Nations FAO. For each country, an estimate is made for the six use types and the biologically productive equivalents of those areas are calculated. These are added together to give a single value of global hectares for each country (the EF), and the results are typically presented in the form of a map (as shown in this chapter) or league table showing how much of the biologically productive land is used for each inhabitant of the countries. The bigger the EF then the greater the area of biologically productive area (land and water) used per capita.

While the above sets out the basic idea behind the EF, one obvious complication is that countries import and export products rather than consume all that they produce. After all, we live in a globalised world where products are traded across national borders. Not all the soybeans produced in the US will be consumed in that country and much of the soybean crop is exported, although, at the time of writing, a trade war with China is seriously impacting upon that flow. The international trade in crop, forest, fish and meat products is significant and needs to be taken account of. Thus, it is important to distinguish between the EF that is based on production within each country, as has been discussed above, and the EF based on consumption within that country. These two can be equated as follows:

$$EF_{Consumption} = EF_{Production} + EF_{Imports} - EF_{Exports}$$

Imports into a country add to its consumption while exports reduce its consumption. A country that exports almost all that it produces will have a relatively low EF for consumption although the EF for production may be high. Similarly, a country that produces little but imports a lot could have a high EF for consumption and a low EF for production. As all the units for EF have been converted to a common unit of global hectares based on biological production then we can perform these additions and subtractions. The key, of course, is having the data of good enough quality available to estimate all the EFs for production, export and import, although as noted above there are extensive and open datasets available from agencies such as the FAO.

The practice of calculating the Ecological Footprint

I have no desire here to swamp the reader with what is a complex and extensive accounting exercise to find the EFs for a country, but here I will provide just one illustration for a single country (Hungary) and single product (apples) that highlights some of the further assumptions that need to be made. Hungary is a landlocked country in Europe (Figure 4.1) with a land area of 93,030 km²; the productive land areas within Hungary are set out in Table 4.2.

The difference between the total area of the country at 93,030 km² and the productive area at 92,123 km² is mostly due to mountains in the north (part of the Carpathians). Most of the land area (70%) is used for crops (including apples) and grazing, while some 22% of the land is covered with forest. The built-up land coverage of 6% includes urban areas as well as land devoted to hydroelectric schemes although these are not particularly prominent in the country. It also needs to be noted that while many of these estimates of land use can readily be made using satellite photography, there is potential for error especially when it comes to built-up land.

Apples comprise just 1 of 164 crop-based products in Hungary and the conversion of crop area and production of apples for 2008 is shown in Figure 4.2. There are three conversions here in total, with the last one being the conversion to global hectares using the equivalence factor of 2.56 (this is used for all crops). The first step in the process, found towards the left-hand side of Figure 4.2,

FIGURE 4.1   Location of Hungary (shaded) in Europe

TABLE 4.2 The productive land areas within Hungary

Land/water use	Area (ha)	% of total area
Cropland	5,552,835	60
Forest	2,045,885	22
Grazing land	877,701	10
Built-up land	560,298	6
Fishing (inland)	175,616	2
Total	**9,212,335**	**100**

Source: Data from Lin et al. (2017).

involves an adjustment of the area of apple production in Hungary (41,100 ha in 2008) to what it should be if the country matched the average yield of apples achieved globally. Some countries achieve higher yields (quantity per land area) of apple than do others, and in 2008 the yield of apples in Hungary was 13.19 tonnes/ha, which was less than the global average of 14.13 tonnes/ha. If Hungary achieved the global average yield, then it would require 40,230 ha rather than 41,100 ha; not a large difference in this case but the adjustment could be more significant for other crops. The conversion of areas based on a global average

hectares   X Yield Factor (YF)   hectares   X Intertemporal Yield Factor (IYF)   hectares   X Equivalence Factor (EQF)   Global hectares

(YF = 0.933 for apple)

(IYF = 0.99 for apple)

(EQF = 2.56 for apple)

Recorded crop area of apples in Hungary for 2008 = **43,100 hectares** (ha) (one square in the diagram above represents 10,000 ha).

The crop area of 41,100 ha produced **568,600 tonnes** of apples in 2008.

This equates to a yield of **13.19 tonnes of apple per ha in a year.**

One ha = 10,000 square metres or the area covered by a square 100 metre long by 100 metres wide.

Area of apple required in Hungary to produce 568,600 tonnes if the apple yield was the same as the global average yield.

Apple yield in Hungary = 13.19 *tonnes/ha*

Global apple yield = 14.13 *tonnes/ha*

Therefore, the apple yield in Hungary is less than the global yield – it is in fact approximately 93% of the global yield in 2008.

If Hungary could boost its yield to match the global yield of 14.13 tonnes/ha then 40,230 ha of the crop would be required rather than the actual area of 43,100 ha.

An adjustment made to the area of apple required in Hungary based on the average world yield of apple.

The adjustment – very minor in this case (99%) -allows for fluctuation in global apple yield over time.

Global bioproductive area equating to the crop land area required for Hungary to produce 568,600 tonnes of apples at the global average yield. It is the *Ecological Footprint for production.*

In this case, the global bioproductive area (Ecological Footprint) equating to apple production in Hungary is **101,958 global ha (g ha).**

The EQF of **2.56** reflects the fact that crop land is generally more productive than the average biological production per unit area of land and water on the planet.

FIGURE 4.2   Calculation of the ecological footprint (production) for apples in Hungary

Source: Own creation based on data in Lin et al. (2017).

yield has been much criticised with regard to the EF, especially as yields for some products can vary greatly.

The second step in the calculation of the EF for apples is a relatively minor one and only applies to cropland (and by extension – built-up land). It adjusts the crop area to allow for variation over time, and it is only really possible to do this for crop-based products given that they have detailed time-series data. Indeed, the adjustment is not large and involves multiplication by an 'Intertemporal Yield Factor', which is assumed to be 0.99 for all crops and built-up land. In the case of apples in Hungary, multiplying by 0.99 means that the crop area is reduced from 40,230 ha to 39,828 ha (once rounded up). Again, much depends on the choice of the Intertemporal Yield Factor.

The third step in Figure 4.2 involves the conversion to global hectares by using the equivalence factor, in this case 2.56. It generates the EF of production for apples in Hungary of 101,958 gha; a footprint of biological productivity much greater than the figure of 41,100 ha we started with as the crop area of apples in that country.

We now have to repeat this for all the other crop-based products and the EF for production comes to 10,986,746 gha. To find the EF for consumption of crop-based products we need to take into account the exports and imports of these products, including crop products that have been fed to livestock and fish, as shown in Figure 4.3. The final EF consumption figure for crops in Hungary comes to 11,474,531 gha. Repeating this process for all land- and water-use types (forest, fisheries, grazing, built-up and carbon uptake) we arrive at the EF for production and EF for consumption in Hungary as set out in Table 4.3.

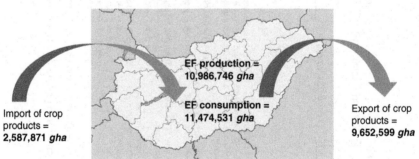

Ecological Footprint for apple production in Hungary = **101,958 *gha***
Process repeated for 164 crop products produced in Hungary and summed gives a total Ecological Footprint for crop production in that country of **10,986,746 *gha***

EF production = 10,986,746 *gha*

EF consumption = 11,474,531 *gha*

Import of crop products = **2,587,871 *gha***

Export of crop products = **9,652,599 *gha***

FIGURE 4.3 Calculation of the ecological footprint (production) for all crops in Hungary

Source: Own creation based on data in Lin et al. (2017).

TABLE 4.3 The ecological footprint components for Hungary (gha)

Use type	EF production	EF imports	EF exports	EF consumption
Crop	18,539,259	2,587,871	9,652,599	11,474,531
Grazing	236,673	701,216	324,112	613,777
Forest products	2,891,020	3,336,679	1,581,903	4,645,796
Fish	21,457	172,762	28,371	165,848
Built-up	1,864,292	0	0	1,864,292
Carbon	13,581,789	14,055,787	11,816,940	15,820,636
Total	**37,134,490**	**20,854,315**	**23,403,925**	**34,584,880**

In this table: $EF_{Consumption} = EF_{Production} + EF_{Imports} - EF_{Exports}$

Source: Data from Lin et al. (2017).

Both the EF production and EF consumption figures for Hungary are largely made up of cropland and carbon uptake, as can be seen in the pie charts in Figures 4.4a and b. Together these comprise more than two-thirds of the total figures for EF production and EF consumption.

But, of course, all the EF figures are totals for the country and much can depend on population size. Just as with the GDP, higher values of the EF could theoretically be obtained by just having more people in a country, but that does not help with country-to-country comparisons. However, as with GDP, it is easy to adjust for this and the EF figures for production and consumption are usually divided by population size in order to provide EF per capita and thereby allow for fairer international comparisons.

The creators of the EF also provide figures for each country of what they refer to as 'biocapacity'. They define this as:

> Biocapacity refers to the amount of biologically productive land and water areas available within the boundaries of a given country.
>
> *(Lin et al., 2017, p. 49)*

This physical limitation means that biocapacity does not have any export or import components and is based solely on the land- and water-use types within the country's borders. Carbon uptake is not included in biocapacity, as within each country's borders this is assumed to occur largely via forestland. Also, perhaps rather oddly, they include built-up land as part of biocapacity of the country. It is hard to imagine how built-up land contributes to biocapacity, but their rationale for this is that "though built-up land does not generate resources, buildings and infrastructure occupy the biocapacity of the land they cover" (Lin et al., 2017, p. 49). Hence, biocapacity can be thought of as the potential or capacity of the country to generate products desired by people and to absorb waste, in this case carbon dioxide primarily

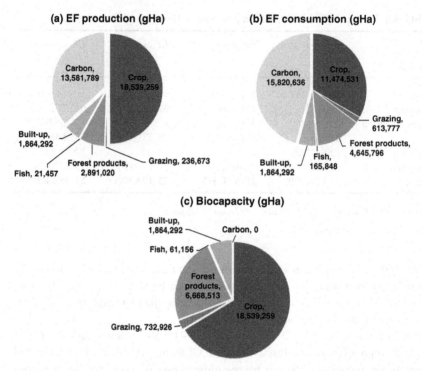

(a) EF production (gHa)

Carbon, 13,581,789
Crop, 18,539,259
Built-up, 1,864,292
Fish, 21,457
Forest products, 2,891,020
Grazing, 236,673

(b) EF consumption (gHa)

Carbon, 15,820,636
Crop, 11,474,531
Grazing, 613,777
Forest products, 4,645,796
Fish, 165,848
Built-up, 1,864,292

(c) Biocapacity (gHa)

Built-up, 1,864,292
Carbon, 0
Fish, 61,156
Forest products, 6,668,513
Crop, 18,539,259
Grazing, 732,926

FIGURE 4.4 Proportional areas of EF production, EF consumption and biocapacity (all in gha) for Hungary

Source: Own creation using data from Lin et al. (2017).

via forestland. Biocapacity is reported in terms of global hectares, following the same logic as that set out above for ecological footprint. The biocapacity figures for Hungary (in gha) are shown in Table 4.4, along with the estimated areas for comparison.

Cropland dominates the biocapacity for Hungary, as shown in Figure 4.4c, and this is perhaps to be expected as carbon uptake is not included as such within biocapacity. Cropland makes up two-thirds of the total biocapacity and is followed by forest products and built-up land.

Biocapacity for cropland (and by extension built-up land) is the same as the EF for production, but some of the other biocapacity figures are different from the EF for production. Indeed, for forest, grazing and fishing the EF for production is less than half of the biocapacity of Hungary. But given that some two-thirds of the biocapacity of Hungary is contributed by cropland, and that this land-use type (along with built-up land) is the same as the EF for production, then the EF production (not including the carbon uptake component) is around 85% of the country's biocapacity and the EF consumption (again without carbon uptake) is 67% of the country's biocapacity.

TABLE 4.4 Ecological Footprint and biocapacity for Hungary

Land/water use	Area (ha)	%	Biocapacity (global ha)	%	EF Production (global ha)	EF production as a % of biocapacity
Cropland	5,552,835	60	18,539,259	66	18,539,259	100
Forest	2,045,885	22	6,668,513	24	2,891,020	43
Grazing land	877,701	10	732,926	3	236,673	32
Built-up land	557,571	6	1,861,565	7	1,864,292★	100
Hydroelectric	2,727	0	2,727	0	–	–
Fishing (inland)	175,616	2	61,156	0	21,457	35
Total	**9,212,335**	**100**	**27,866,146**	**100**		

Note: ★The EF of production for built-up land includes hydroelectric.

Source: Data from Lin et al. (2017).

Critique of the ecological footprint

Despite all the work that goes into finding the EF, and the reader can get a taste of that from the previous section, it is worth pointing out that it is something of a disputed index, with very much what we in the UK call a 'Marmite' (a type of foodstuff flavouring) feel to it; some love it while others certainly do not! There are many assumptions, and if we change any of them the value of the EF will also change. Perhaps understandably, there have been many debates about these in the academic literature. Given the complexity of calculating the EF, it seems reasonable to assume that consumers of the EF, many of whom will not be specialists in this field, may tend to take the EF figures as givens and not delve into or question the details of how they were reached. Even the example for one crop in one country (apples in Hungary) shows how involved the calculation can be. Thus, it seems likely that complex indices such as the EF may often be taken at face value. The reader interested in exploring the critiques of the EF and how its creators respond can find them in Chapters 16 and 33 of Bell and Morse (2018).

But for all the faults of the EF claimed by its critics, it does have a resonance with a lay audience that many other indices arguably struggle to find. The higher the value of the ecological footprint – the bigger the footprint – then the greater the impact on the planet, and given that EF is calculated for each country we can get a feel for which countries have the biggest impact. Compare this with, for example, the Environmental Performance Index (EPI) and its scale of 0 to 100. Sure, we can find a value of the EPI for a country and report that and the country's ranking relative to others, but the index has no inherent meaning for us; it is just a number within an artificial scale. The EF is different as it is expressed in terms of something we can see and visualise – an area – and this in turn chimes with our senses of fairness; just why should some countries take so much more of the Earth's finite resources than others? The EF speaks to our souls as it equates in the minds of many to senses of greed and exploitation.

The global pictures of the ecological footprint

The global pictures in terms of the EF production and EF consumption, adjusted to a per capita basis, are set out in Figure 4.5. They appear to be similar. For the EF production, the values are highest in North America, Australia, New Zealand, northern Europe and North Asia, while values are lower in Africa, parts of South America and South Asia. With regard to EF consumption, the picture is not too dissimilar, with relatively low values in Africa, South America and South Asia, and higher values in North America, Northern Europe and Northern Asia. Saudi Arabia, a kingdom with a surface that is largely comprised of desert, stands out on both maps as having higher EF production and high EF consumption. A comparison of these two versions of EF by means of a scatter-plot is shown in Figure 4.6, with each dot representing a single country. Notice how for lowish values of both these versions of the EF there is good alignment, but as the values increase then so does the scatter between them. It seems that at higher values of the EF the link between production and consumption breaks down, and this can be readily explained by importation. Countries with high consumption tend to import much of what they consume and thus for countries with high EFs for consumption the simple linear link with production is likely to disappear. Hence, countries with high EF for consumption are 'sucking' in production EF from elsewhere.

One of the key aspects of the EF that always seems to jump out is the difference in consumption between countries that have broadly similar living conditions; for example, the US compared with Europe. The comparison of indices is something we will return to in Chapter 9, but here it only needs to be noted that the EF can depend a great deal on the efficiency of use of resources and high values for the EF are not a necessity for a good quality of life. Indeed, one of the perverse interpretations of the EF is that high values are good because they equate in some way to the development of a country. After all, industrialisation must surely mean that resource use increases? Thus, a large EF may unfortunately be regarded as a badge of honour – equating to progress or perhaps even to a sense of pride that 'our country' is having such a large impact on the globe.

The view of the world through the lens of biocapacity yields a surprisingly uniform picture (Figure 4.7). We don't see the major differences that we do with the EF, and the few 'outlier' countries (dark shading on the map) are relatively small in terms of land area. Remember that with biocapacity we do not have any import or export components. The three countries that stand out with very high biocapacity are Suriname, Gabon and Guyana: Countries that, while relatively small in terms of total area, do have very high percentage forest cover; for Gabon and Guyana it is 77% and for Suriname, the country with the highest biocapacity per capita, it is 94%. It is these very high levels of forest cover that generate the high biocapacity of those countries. Other than these outlier countries with high forest cover, the world looks like a relatively uniform place in terms of biocapacity. It is when we look at the impact on our planet that we see the big differences.

(a) Ecological Footprint (production)

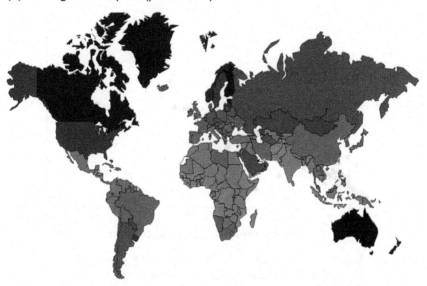

(b) Ecological Footprint (consumption)

FIGURE 4.5 The two variants of the ecological footprint. Darker areas equate to higher values. These data are for 2013 but published in 2017. (a) Ecological footprint (production), (b) ecological footprint (consumption)

Source: Own creation based on data from 2017 Global Footprint Network. National Footprint Accounts, 2017 edition. Contact Global Footprint Network at data@footprintnetwork.org for more information.

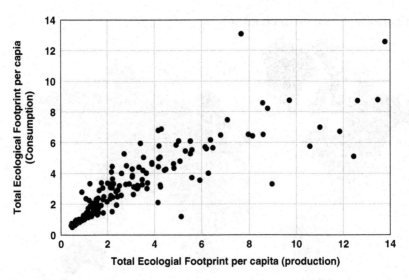

FIGURE 4.6 Relationship between EF consumption and EF production for countries. Data are from 2013 and each dot represents a single country

Source: Own creation based on data from 2017 Global Footprint Network. National Footprint Accounts, 2017 edition. Contact Global Footprint Network at data@footprintnetwork.org for more information.

FIGURE 4.7 Biocapacity for countries. Darker areas equate to higher values. In terms of biocapacity, the world looks like a remarkably uniform place

Source: Own creation based on data from 2017 Global Footprint Network. National Footprint Accounts, 2017 Edition. Contact Global Footprint Network at data@footprintnetwork.org for more information.

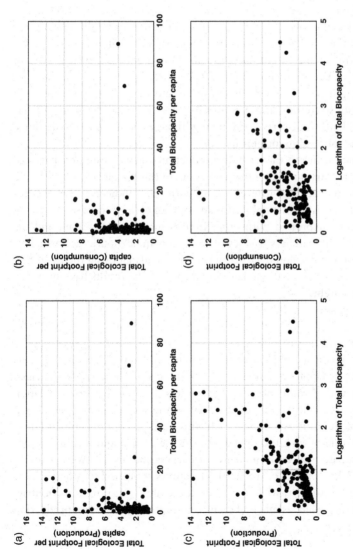

FIGURE 4.8 The ecological footprint plotted against biocapacity. All data are from 2013. The graphs in the top row show the raw data. The result is a bunching of countries towards the left-hand side. The graphs on the bottom row use the logarithm of biocapacity to stretch out lower values along the horizontal axis. Even so, there is no obvious relationship between the EF and biocapacity

Source: Own creation based on data from 2017 Global Footprint Network. National Footprint Accounts, 2017 edition. Contact Global Footprint Network at data@footprintnetwork.org for more information.

How do EF production and EF consumption relate to biocapacity? As noted above, biocapacity represents the capacity of countries to generate products desired by people and to absorb waste, and the boundary is very much that of the national state; there are no adjustments for import or export. Graphs of the EF (production and consumption) against biocapacity (Figures 4.8a and b) generate a strong bunching effect towards the left-hand side of the graph where biocapacity is between 0 and 20 gha, but even here there is a wide range of the EF. Given this bunching effect, it can be hard to discern any pattern, but, as noted in Chapter 3 with the GDP/capita component of the HDI, it is possible to spread out the bunching of biocapacity values at the low end of the scale and compress them at the higher end of the scale and thereby reduce the gap between the largest and smallest values by taking the logarithm. The creators of the HDI do this in order to avoid dominance of GDP/capita in the HDI given that the gap between the largest and smallest values is so wide. We will return to this technique of taking logarithms in Chapter 9 so I will not discuss it in detail here, but the results of taking the logarithm of biocapacity are shown in Figures 4.8c and d). It has to be said that, even with this trick of simultaneously stretching and compressing the horizontal axis using logarithms, it is still hard to discern a convincing linkage and there is a wide range in the EF for any one value for biocapacity.

It is the variation in the EF for both production and consumption between countries that is the main message that comes out of this scatterplot exercise, and in Chapter 9 we will see how this compares with other indices covered in this book. Indeed, the EF is often employed as an estimate of the impact on the planet of human benefits, expressed with other indices such as GDP/capita and the HDI. Hence many researchers have estimated gains in economic performance, happiness and human development per impact on the planet using the EF as that measure of impact. As we will see later in Chapter 9, the results have provided much food for thought.

Conclusion

The EF is undoubtedly a complex index, as can be seen with the example of one crop in one country (apples in Hungary), but it is one that has arguably managed to achieve a degree of resonance with consumers of the index via its relation to a limited resource that we can all appreciate – the surface area of the planet. Unlike many of the unit-less indicators in this book, which are just numbers on an arbitrary scale of 0 to 100 or 0 to 10, the use of area and its implications about greed resonate well with our sense of fairness and justice. Why is it that some countries take so much of the Earth's resources compared to others? What gives them the right to do that? It is not as if those high-consumption countries produce all that they need, as much of their consumption comes from production in other countries.

References

Bell, S and Morse, S. (eds.) (2018). *Routledge Handbook of Sustainability Indicators and Indices*. Routledge, London.

Gliessman, S R (2007). *Agroecology. The Ecology of Sustainable Food Systems*. 2nd edition. CRC Press, Boca Raton.

Lin, D, Hanscom, L, Martindill, J, Borucke, M, Cohen, L, Galli, A, Lazarus, E, Zokai, G, Iha, K, Eaton, D and Wackernagel, M (2017). *Working Guidebook to the National Footprint Accounts*. Global Footprint Network, Oakland.

Liu, M, Li, W, Zahng, D and Su, N (2015). The calculation of equivalence factor for ecological footprints in China: A methodological note. *Frontiers of Environmental Science and Engineering* 9 (6), 1015–1024.

Rees, W and Wackernagel, M (1996). *Our Ecological Footprint*. New Society Publishers, Gabriola Island, British Columbia.

Wright, L, Kemp, S and Williams, I (2011). Carbon footprinting: Towards a universally accepted definition. *Carbon Management* 2 (1), 61–72.

Further reading

Berners-Lee, M (2011). *How Bad Are Bananas? The Carbon Footprint of Everything*. Profile Books, London.

Franchetti, M Jand Apul, D(2012). *Carbon Footprint Analysis: Concepts, Methods, Implementation, and Case Studies*. CRC Press, New York and London.

Muthu, S S (ed.) (2016). *The Carbon Footprint Handbook*. CRC Press, New York and London.

5

ENVIRONMENTAL PERFORMANCE INDEX

Introduction

At the great risk of stating the obvious, we live on a planet called Earth. At the time of writing this book there are no human beings living anywhere else, with the small exception of a handful of astronauts inhabiting a space station orbiting our planet. This world has to provide us with everything that we need, although the vast bulk of our energy comes from the sun in one form or another. All this may change in the future, of course, as people begin to explore the other planets and moons of the solar system, and at some time there may well be colonies inhabiting those places, but for the foreseeable future we are limited to living on one planet. This is our home and it supports our lives as well as those who lived before us and future generations. The Earth has an impact upon us and we have an impact upon it, as evidenced by a number of changes witnessed over the past few hundred years, from the local (e.g. pollution events in rivers) to the global (human-induced climate change). Unsurprisingly there have been various efforts to develop indicators and indices to asses our impact upon the Earth, and the ecological footprint (EF) is one of them. But the EF is an index that, at its heart, focusses upon resource consumption; it is designed to capture what we take out of the planet and expresses this in terms of a 'footprint' – the surface area required to provide those resources. There are other indicators and indices that seek to explore impacts on the planet, and indeed us, that arise from changes we cause to the environment. Pollution to the land, air and water and subsequent impacts on the environment and human health are all examples of this, and they are not captured directly in the EF. It can be assumed, of course, that pollution can reduce crop yield and forest products, to name but two, but the EF does not attempt to assess pollution or indeed the quality of our environment.

In this chapter we will explore a tale of two indices that were designed to assess, in part, the quality of our environment and our impact upon it: The Environmental Performance Index (EPI) and its predecessor, the Environmental Sustainability Index (ESI).[1] It is a rich story that spans 18 years, and one that witnessed a significant and intriguing change of emphasis as the ESI gave way to the EPI. As with all the indices in this book the back story behind the ESI/ EPI provides some interesting insights into how we try to represent the world we live in. Both the ESI and EPI are complex indices with many components all weighted differently and, as we have seen with the Human Development Index (HDI), in some cases raw data were transformed to limit an excessive influence. The assumptions that underpin both the indices are numerous, although, in fairness and in common with almost all the indices discussed in this book, the custodians of the indices have been very transparent about what they did and why.

Capturing sustainability with an index

The ESI was developed in the late 1990s and first published as a pilot study in 2000. It was commissioned by the World Economic Forum (WEF), a meeting of large companies, pressure groups, governments and international agencies held each year in Davos, Switzerland, and implemented by Yale and Columbia universities in the US. The idea was no different from that of the other indices presented in this book: To have a headline index that would draw attention to the cause, in this case environmental sustainability, and seek to bring about improvements by transparency and peer group pressure. Countries would be able to see where they were positioned in the ESI league table and compare themselves with countries they saw as their peers. Poor-performing countries would feel pressure to improve their performance in time for the next league table ranking.

The ESI was published again in 2001, 2002 and 2005. The gap between 2002 and 2005 reflected a view among the creators of the index that it was not necessary to publish the ESI every year. Unlike the economy, which can see rapid changes in a few months, or indeed even quicker than that, it was assumed that environmental sustainability changes over longer timescales. Also, of course, there was the usual concern regarding data availability.

The ESI is an aggregated index along the lines of others presented in this book, but here the aggregation results from combining many datasets that encompass highly diverse aspects of environmental sustainability, from ambient pollution and emissions of pollutants to impacts on human health and being a signatory to international agreements. As far as indices go, the creation of such a multidimensional and multifaceted index as the ESI certainly reflected a great deal of ambition, and it made the HDI with its three elements look highly simplified by comparison.

Values of the ESI for each country vary between 0 (most unsustainable) to 100 (most sustainable), and, in common with all the indices in this book, the results for each country are published in the form of a league table with associated colour-shaded maps. Given the number of components and their diversity, it is unsurprising that the ESI methodology is complex, and the details do not need to be presented here. In essence, the calculation of the ESI is an aggregation process that starts with variable (raw datasets) that are combined into 'indicators', which in turn are placed into one of five components (or themes) and then averaged to generate the ESI as shown in Figure 5.1.

An indicator may comprise more than one variable and the themes also differ in terms of the number of indicators that comprise them. To add to the complexity, the number as well as nature of the variables and indicators included in the ESI varied significantly over the various releases of the index, and in that sense the ESI is arguably one of the diverse indices in this book, but, as we shall see in Chapter 10, not the most diverse. However, the five components (or themes) did remain consistent over the years, even though the names given to them varied slightly. They are broadly:

- **Environmental systems**: A sort of snapshot of the quality of the environmental system. It includes, for example, the concentration of pollutants in the environment.
- **Environmental stresses**: This component comprises a set of indicators that relate to pressure on the environmental system. For example, this component includes the rate of release of pollutants into the system, the rate of deforestation and human population growth.
- **Human vulnerability**: This component covers aspects of human health and welfare impacted upon by the environment. It includes, for example, the incidence of human diseases, malnutrition, access to drinking water and the rate of infant mortality.

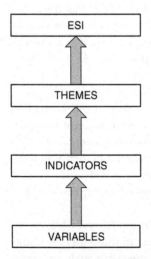

FIGURE 5.1 The hierarchy from variables to the Environmental Sustainability Index

- **Social and institutional capacity**: This is a large component, the largest of all five in terms of number of variables included. The focus is on country capacity to look after the environment, and it spans aspects such as investment in and outputs from research, the presence/extent of protected areas, and energy efficiency. It also includes arguably more nebulous variables such as civil liberties.
- **Global stewardship**: This component is focussed on membership/engagement/compliance with various environmental international agreements as well as the release of pollutants such as carbon dioxide across national borders.

The five themes do have a logic in terms of assessing sustainability. They combine concerns for the environment along with more social and economic concerns, and also include the capacity involved in trying to achieve sustainability. Given that definitions of what we mean by sustainability can themselves be rather all-encompassing then the breadth in the list above is unsurprising. For example, here is perhaps the most commonly reported definition of sustainable development produced by the World Commission on the Environment and Development (WCED) back in 1987:

> Development that meets the needs of current generations without compromising the ability of future generations to meet their needs and aspirations.
>
> *(WCED, 1987, p. 43)*

What is meant by 'needs' let alone 'aspirations' is not defined with any precision in the WCED report. It can be implied that these relate to an ability to lead a healthy and fulfilling life, and that includes, of course, a requirement for future generations to also have access to resources that they will need "to meet their needs and aspirations", even if we may not know what they are. The environment matters in all this as it provides resources (food, water, medicines, oxygen, etc.) needed to allow us to lead a healthy and fulfilling life. Disrupt that environment and we, or our future generations, will pay the consequences. The ecological footprint (Chapter 4) is clearly of relevance here, as it assesses the pressure we are placing on the planet's resources to meet "our needs and aspirations", but, although important, there is much more to sustainable development than that. Indeed, we often conceptualise the essence of the WCED definition above into a diagram comprising three overlapping circles, as shown in Figure 5.2. The three circles represent the economic, social and environment circles within which we live, and sustainable development is represented by the area where they overlap.

The diagram in Figure 5.2 is not favoured by everyone, it has to be said, partly because the circles can be thought of as representing our level of interest or concern, and in that case could well end up being relatively different in size. As we have already seen in Chapters 2 and 3, the economic sphere tends to dominate the thinking of many. Also note that the definition above is for 'sustainable

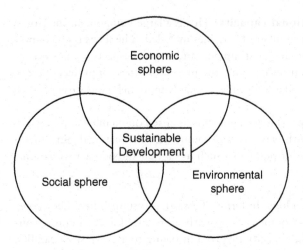

FIGURE 5.2 The interactions between environmental, economic and social spheres that form sustainable development. The central area, where all three circles overlap, is the space represented by sustainable development

development' rather than 'sustainability'. Some see these terms as being synonymous while others argue that they are different. Hence, those belonging to the latter school of thought would argue that we can apply the essence of the definition above to the separate components of sustainable development in Figure 5.2. This allows us to think in terms of separate concepts we can call 'environmental sustainability', 'social sustainability' and 'economic sustainability'. The definition of 'environmental sustainability', the term that provides the title for the ESI, can be defined as:

> A state in which the demands placed on the environment can be met without reducing its capacity to allow all people to live well, now and in the future.
>
> *(Financial Times Lexicon, available at lexicon.ft.com/ Term?term=environmental-sustainability)*

The heart of this definition looks very similar to the WCED one above for sustainable development, although it is framed more narrowly in terms of demands and capacity. Even so we still have the inclusion of people and their ability to "live well". Therefore, it is surely axiomatic that the definition for environmental sustainability has to exist within the WCED definition, and thus it can seem odd to have them expressed separately. It can seem somewhat reductionist to separate out the environmental, social and economic dimensions of sustainable development and treat them separately when one is inevitably forced back from that narrow position to consider the whole. Anyway, I appreciate that the head can start to throb at this point so I do not want to pursue this argument any further. Needless to say, the composition of the ESI has resonance with both of the definitions above – broad and narrow – and this is not surprising.

While the list of themes has a logic, it is indeed an ambitious task to try to capture sustainability in a single index. There have been many efforts over the years to develop lists of indicators with each one designed to capture one component of sustainability, and these are often presented in a single table or perhaps a diagram, but few have attempted to combine all the information into a single index. If the reader is interested in such sustainability indicator frameworks, I can refer them to a number of books I co-authored with a colleague, Simon Bell, in the late 1990s and early 2000s (Bell and Morse, 1999, 2003). The number of indicators and variables within the ESI themes has varied over the years and the distribution has never been evenly balanced between them. In Figure 5.3 I have presented the number of variables and indicators included in the various versions of the ESI from 2000 (the

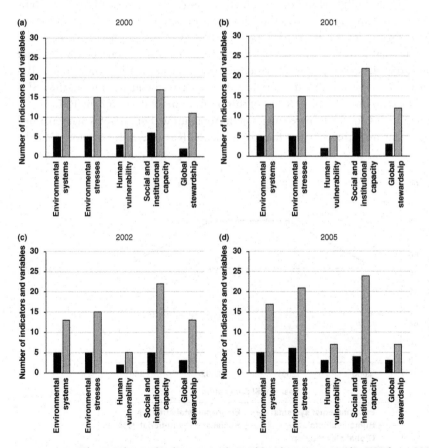

FIGURE 5.3 The numbers of indicators and variables that comprise the ESI from 2000 (the pilot study) to 2005. Darker shaded bars show the number of indicators and the lighter shaded bars show the number of variables

Source: Own creation using data from the ESI reports (2000 to 2005).

pilot) to the final release in 2005. Across the foot of each graph you can see the five components, with one column (darker shaded bars) for the number of indicators and one (lighter shaded bars) for the number of variables. The number of variables is greater than the number of indicators, as an indicator can comprise a number of variables. From the series of graphs in Figure 5.3 it can be seen that the total number of variables in the ESI ranged between 65 (in 2000) and 76 (2005); way ahead of many of the other indices covered in this book although, arguably, the ecological footprint is also just as 'data hungry'.

Interestingly the last two themes in the list above, 'Social and institutional capacity' and 'Global stewardship', had a significant proportion of the total number of variables in each of the releases of the index. In 2001 and 2002 these two components had over half the total number of variables, while in 2000 (the pilot release) they had 43% of the variables and in 2005 the figure was 41%. However, the ESI is found by taking the arithmetic average of the values for the five components and any imbalance in the number of variables within each component is not compensated for in the final value of the ESI. Thus, just because 'Social and institutional capacity' and 'Global stewardship' have more variables it does not mean that those components have a greater weight in the ESI – they do not. But what it does mean is that the variables do not have an equal weight in the ESI. I will pick up on the ramifications of this point in more detail for the EPI.

Figure 5.4 shows the percentages of the total number of variables that remained consistent over time for the five components. It is interesting to note that the

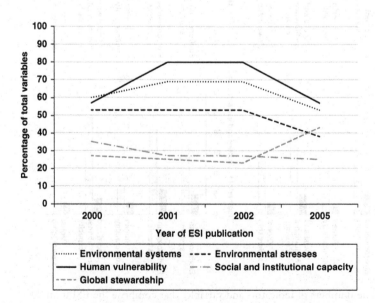

FIGURE 5.4 Consistency in variables over time within the five components of the Environmental Sustainability Index

Source: Own creation based on data in the ESI reports of 2000, 2001, 2002 and 2005.

creators of the ESI were more consistent about the variables they included for the 'Environmental systems', 'Environmental stresses' and 'Human vulnerability' components compared with the other two. These three components typically had more than 50% of their variables remaining the same from 2000 to 2005, while 'Social and institutional capacity' and 'Global stewardship' had between 23% and 35% of their variables remaining the same. This is probably understandable given that the 'Environmental systems', 'Environmental stresses' and 'Human vulnerability' components had many variables (e.g. those related to water and air pollution) that can be considered as relatively 'standard' and well-established in environmental science and human well-being. Hence the choice of variables in these themes is to an extent set by an established precedence within the scientific community. However, the 'Social and institutional capacity' and 'Global stewardship' components are arguably less well-established and so a degree of variation over what to include may be anticipated.

The ESI was released each year accompanied by attractively presented, lengthy and detailed reports that set out all the details of the calculations, including the assumptions that were taken regarding weighting of the variables within indicators and themes, transformations, capping of variables, etc. The reports are certainly extensive in terms of covering the technical details of the ESI, but given the numbers of indicators and variables and the complex ways in which they are adjusted and combined it is hard to imagine the casual consumer of the index being able to spend much time disentangling the construction process. Therefore, it could be argued that the 'transparency' while commendable, may not matter that much in reality for the majority of consumers of the ESI. Nonetheless, as one of the founders of the ESI has pointed out:

> The true value [of the ESI] is the ability to break down the score on an issue by issue basis.—Daniel Esty
>
> *(New Scientist, 2005, p. 6)*

Thus, the creators of the ESI were open about the values allocated to the themes and indicators meaning that anyone could see how they compared to each other within and between countries. It has to be noted that all the indices covered in this book are released with details about their construction so anyone, if they so wish, can disentangle them and work out for themselves the process and assumptions. This information may be in a separate 'technical' annex or perhaps included within spreadsheets, but the information is there and available. Thus, anyone should be able to break down the HDI or the ecological footprint to explore them on an "issue by issue basis", although the process is more challenging for some indices than it is for others – and the ESI is arguably at the higher end of the complexity spectrum. However, in reality it seems reasonable to assume that most of the intended consumers of the indices will simply opt to accept the published value of the index and trust the decisions of those who have constructed them.

The ESI has received a degree of criticism since its first release and much of it can be summed up in a paper published in *The Ecologist* magazine in 2001. Given the large number of variables in the ESI and how these eventually 'morph' into the ESI, it is unsurprising that many of the criticisms revolve around assumptions such as availability of quality data, how data gaps are filled, transformations and weightings. For example, at the highest level of construction of the ESI – the step from the five components to the ESI – there is a key assumption that each component is equally weighted, so the ESI is the arithmetic average of the five. This is not an unusual assumption and many indices with multiple components also assume equal weightings using the arithmetic mean or maybe the geometric mean, as for the more recent versions of the HDI. But two of the components in particular – 'Social and institutional capacity' and 'Global stewardship' – seem to have a strong link with economic wealth as we will see later in this chapter. These components have many variables that reflect expenditure, such as investment in research and membership of international agencies, and wealthier countries would be expected to do better here. In addition, the 'Human vulnerability' category, which includes variables such as disease incidence, access to water, and malnutrition, would also seem to have an inherent bias in favour of wealthier countries, where households can afford to purchase food of adequate quantity and quality as well as medicines, and can afford high quality sanitation and healthcare systems. We can argue, of course, as to whether all this is important as, after all, if some countries do have more money than others then what matters, and what is captured in the ESI, is what they do with it. Just because a country is relatively wealthy it does not necessarily mean that its government will invest in the variables used in the ESI, but, whichever way you look at it, countries that are poorer will almost certainly be less able to do as well as wealthier ones. It seems reasonable to assume that changing the balance of the components in terms of their influence on the ESI would potentially change the outcomes a great deal. That is exactly what the authors of the *Ecologist* article did; they simply reweighted the ESI components to allow for disparities in national wealth and, as a result, the country league table based on the ESI changed dramatically.

Environmental Sustainability Index to Environmental Performance Index

The last version of the ESI was published in 2005; it was replaced in 2006 by a new index created by the same two universities: The Environmental Performance Index. The EPI was certainly a more constrained index than the ESI in that it focussed entirely on environmental elements rather than trying to span the social and economic considerations that were in the ESI, and the number of variables was greatly reduced as a result. The creators of the index also dropped the term 'sustainability' and instead opted for 'performance', and, given the brief debate in the previous section regarding the meaning of sustainability, this shift to performanc' also has a feel of making things simpler. The EPI has certainly been more long-lived that its predecessor. After its first release in 2006, the EPI has regularly been published every two years – in 2008, 2010, 2012, 2014, 2016 and 2018. This makes seven EPI reports to date, compared to four reports for the ESI (2000, 2001, 2002 and 2005), and the timetable of publishing the EPI every two years does give a feel of robustness relative to the ESI.

The EPI has fewer variables than the ESI, and the aggregation is also much simpler; although the terminology is different. For the EPI, the variables are now referred to as indicators and these are aggregated to produce issues. Issues are further aggregated to yield objectives and then, eventually, the EPI. This is set out in Figure 5.5. Starting from the top, the two objectives are given as 'Ecosystem health' and 'Ecosystem vitality' and, for the EPI 2018, the issues within them are as follows:

- Air quality
- Water quality
- Heavy metals
- Biodiversity and habitat
- Forests
- Fisheries
- Climate and energy
- Air pollution
- Water resources
- Agriculture

At first glance, while the terminology changes between the ESI and EPI do not help, this list does have resonance with the sorts of variables included in the 'Environmental systems' and 'Environmental stresses' components of the ESI, but this is in reality something of a mirage and the point will be returned to later.

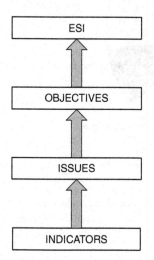

FIGURE 5.5   The hierarchy from indicators for the Environmental Performance Index

The EPI issues are captured by a total of 24 indicators in the EPI of 2018; far less than half the total number of variables employed in all forms of the ESI and also less than the 28 to 38 variables in the 'Environmental systems' and 'Environmental stresses' components of the ESI. The EPI thus represents a significant drop in complexity and greater focus than seen with the ESI. It takes the 'distance to target' approach discussed in Chapter 3 for the Human Development Index, with the targets set by the creators of the EPI. Thus a higher value for the EPI represents a closer match to desired target with regard to environmental performance.

Given the hierarchy of aggregation seen in Figure 5.5, it is perhaps understandable that the EPI has much the same issue of weighting seen for the ESI. Some of the issues are comprised of multiple indicators in much the same way as some indicators of the ESI are comprised of more than one variable. In the EPI the indicators that comprise issues may have different weightings and the same applies to issues that comprise objectives. The result of all this is that the relative contribution of indicators in the EPI towards the final value can be markedly different. In the pie chart shown in Figure 5.6 we can see all 24 indicators

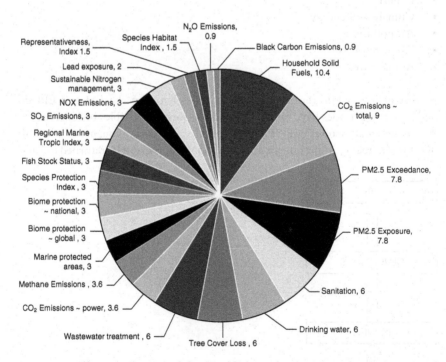

FIGURE 5.6 The contribution of the 24 indicators that comprise the Environmental Performance Index of 2018 towards the final value of the index. Numbers are the percentage contribution to the EPI, with the four main ones in bold text

Source: Own creation based on data in the EPI report of 2018.

that comprise the EPI 2018 and I have added their contribution, in percentage terms, towards the final EPI once all the weightings throughout the hierarchy of aggregation have been taken into account. One of the indicators, 'Household solid fuels' comprises just over 10% of the final value of the EPI, while N_2O and 'Black carbon' (e.g. soot) emissions account for less than 1%. This difference in percentage contribution might not look like much, but what we see in Figure 5.6 represents a relatively large range of weights and it is not immediately apparent why the weightings should be as they are. Indeed, one could imagine a wealth of different weights if a range of experts were asked to carry out this process and it is not inconceivable that weight could depend on expertise. 'Fish stock status' and 'Marine protected areas', for example, only have a weight of 3% in the EPI of 2018 but I could well imagine a fishery expert making a convincing case to increase those values. As objective and scientific as the EPI may look it clearly has much subjectivity at its heart.

The world as seen with the ESI and EPI

How does the world look through the respective lenses of the ESI and the EPI?

Figure 5.7 shows the world as it would appear if we could see the values of the ESI for nation states. There are two versions of the ESI presented here, one for 2001 (graph a) and the other for 2005 (graph b). Given that the difference between them is only four years you may not have expected to see much change, and indeed there is not. A cursory glance at the global pictures from those two years does present much similarity, and we can elaborate further on this by using a simple scatter graph of ESI 2005 plotted against ESI 2001, as shown in Figure 5.8. Four years may seem like a long time in our lives, and it is, but in terms of many of the components captured within the ESI they do not change that fast. It is also interesting to see the data gaps in 2005, the countries coloured white, perhaps suggesting that the custodians of the ESI were struggling to get quality data for all countries. Given the complexity of the ESI this is perhaps understandable: Complex indices with many components will generate a heavy demand for data and gaps do seem to be more likely.

In both these years (2001 and 2005) the overwhelming picture presented by the ESI is one of a band of unsustainability stretching from Africa through southern Asia to China. By way of contrast, Europe and North America stand out as beacons of sustainability, as does South America. This may perhaps come as something of a surprise, given that much of industrial production is in Europe and North America, and these places are also major consumers of fossil fuels. Surely these facts would mean that these more developed countries would be major polluters? Africa is still relatively under-industrialised and its energy consumption per capita is also relatively low, and as a result it seems reasonable to suppose that it does not generate much pollution. So how come Africa, in particular, ends up being so unsustainable? Well it all comes down to what is in the ESI. Among its many components are some that have a strong positive link

(a) Environmental Sustainability Index 2001

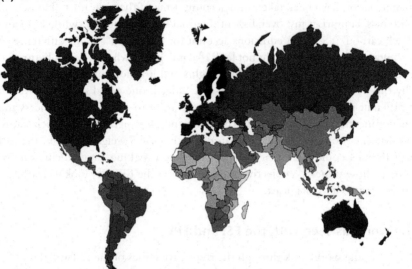

(b) Environmental Sustainability Index 2005

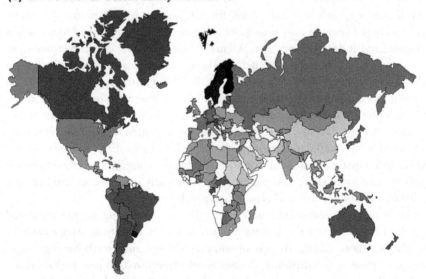

FIGURE 5.7   The world through the lens of the Environmental Sustainability Index
between 2001 and 2005. The darker the shading then the higher the
value (or more environmentally sustainable) the country is deemed to be.
Countries in white have missing values. (a) Environmental Sustainability
Index 2001, (b) Environmental Sustainability Index 2005

Source: Own creation based on data in the ESI reports of 2001 and 2005.

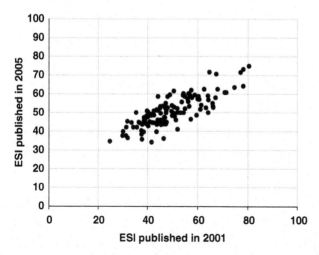

FIGURE 5.8 Relationship between the ESI of 2001 and the ESI of 2005. Each dot represents a single country

Source: Own creation based on data in the ESI reports from 2001 and 2005.

to wealth, and it goes without saying that the richer parts of the world will do well with them. We can illustrate this by looking at the world through the five lenses (components) of the ESI: Environmental systems, Environmental stresses, Human vulnerability, Social and institutional capacity and Global stewardship. The results are shown in the collage of graphs in Figure 5.9 with darker shades representing better performance in terms of their contribution to the ESI, and the reader is invited to compare these with the maps in Chapter 2 where we discussed economic indices. First, looking at (a) Environmental systems and (b) Environmental stresses, where the components have indicators covering pollution, we see that many developed countries such as those of Europe and North America, but also China and India, tend to do worse than countries of Africa and Latin America. This is not unsurprising given that the developed countries tend to have higher levels of industrialisation and hence pollution. Second, looking at (c) Human vulnerability we see a reversal, with darker shades concentrated in the more developed countries and also Latin America and much of Asia. The stand-out continent here is Africa. Given that the Human vulnerability component has indicators that cover human disease, malnutrition and access to water of good quality, it is probably unsurprising to see this pattern. However, the most marked contrasts are perhaps to be seen with the final two components: (d) Social and institutional capacity and (e) Global stewardship. With Social and institutional capacity the better performance is almost entirely concentrated in Europe, North America and Australia. The rest of the world appears to be at the other end of the scale. The contrast is a stark one – one of the largest seen in any of the maps in this book – and clearly resonates with the assumption that performance in this component of the ESI is strongly linked to wealth. With Global stewardship we

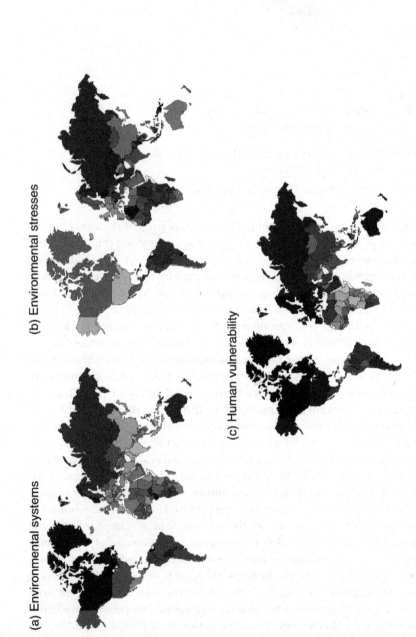

(a) Environmental systems

(b) Environmental stresses

(c) Human vulnerability

FIGURE 5.9 Series of maps showing the five themes of the Environmental Sustainability Index for 2005

Note: In all these maps darker shades equate to higher values and better (more positive) contributions towards environmental sustainability.

Source: Own creation based on data in the ESI report from 2005.

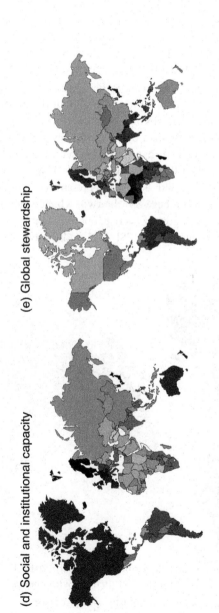

(d) Social and institutional capacity

(e) Global stewardship

FIGURE 5.9   (Continued)

see an almost complete reversal of the picture presented by Social and institutional capacity, maybe because this component includes the transnational spread of pollutants such as carbon dioxide along with being signatories to international environmental agreements. However, once these components are put together one cannot help but feel that the overall result is a masking of components that measure pollution and a story that suggests good environmental sustainability for the wealthier countries. But that is not the whole story.

The shift from the ESI to the EPI, with a reduction in number of indicators and a shift away from variables that were in the 'Social and institutional capacity' component in particular, would seem to have removed many of the components that were directly related to national wealth. Yet the story of the world seen through the eyes of the EPI published in 2018 (Figure 5.10) is not that dissimilar from that presented by the ESI of 2005, 13 years earlier. Again, we are presented with a global picture of a generally good environmental performance by Europe, North America and Australia and a bad environmental performance from the rest of the world. In fact, the global differences between the developed and the developing world appear to be more marked than they were with the ESI 2005. We can see the relationship between the EPI 2018 and the ESI 2005 by looking at a scatterplot between them in Figure 5.11, with each dot representing a single country. It does look like there is a relationship between the two, which is not unexpected given that the components have a degree of overlap, but the points are not as neatly lined up as they were when we plotted the ESI 2001 and the ESI 2005; there is more

FIGURE 5.10 The Environmental Performance Index for 2018. Darker shading indicates higher values for the EPI (better environmental performance)

Source: Own creation based on data in the EPI report of 2018.

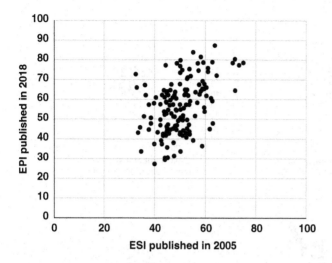

FIGURE 5.11   Relationship between EPI 2018 and ESI 2005. Each dot represents a single country

Source: Own creation using data from the ESI 2005 and EPI 2018 reports.

'messiness'. Also notice how the points spread between the upper-20s and 90 for the EPI but only between the upper-30s and mid-70s for the ESI. In fairness, while there is some overlap between the ESI and the EPI there are also some differences and we would not expect the same sort of clear relationship as we would see when comparing two versions of the ESI. There will be countries that do well in the ESI and not so well with the EPI and vice versa. But the starkness presented by the world represented via the EPI is nonetheless surprising.

Perhaps more surprising, the picture of the EPI 2018 seems to be at odds with the two components of the ESI with which we would expect, at first glance, the EPI to have the closest relationship – Environmental systems and Environmental stresses. In Figures 5.12 (a and b) you can see the EPI 2018 plotted against these two supposedly 'close relative' components of the ESI. With Environmental systems there seems to be no relationship at all, while with Environmental stresses the picture, if anything, is one of a decline in the EPI 2018 as this component increases. Of the other three components of the ESI 2005 (Figure 5.12c, d and e), the EPI 2018 seems to have the best relationship with the two components that, as noted above, arguably have the strongest relationship to national wealth – Human vulnerability' (Figure 5.12c) and Social and institutional capacity (Figure 5.12d). This is an issue I will return to later in Chapter 9 and indeed is a danger behind looking for simple interpretations in scatterplots such as those presented here; the indices being compared may not be directly related but both may be related to a common underlying variable.

Nonetheless, we are left with the question as to whether wealthier countries really are that good at managing their environment compared to less wealthy

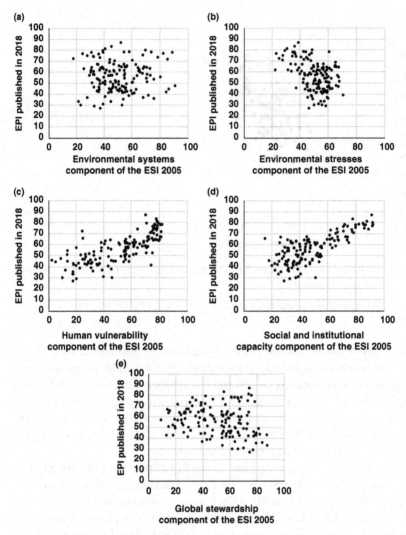

FIGURE 5.12 The Environmental Performance Index (2018) plotted against the five themes of the Environmental Sustainability Index of 2005. Each dot represents a single country

Source: Own creation using data from the ESI report of 2005 and the EPI report of 2018.

countries, as suggested by the global picture through the lens of the EPI. Well, again much depends upon what is included in the EPI just as it does with the ESI. While the EPI is simpler (in terms of the number of its components) and arguably more focussed, the story being told is still dominated by the indicators that have been selected and how they have been manipulated to generate the index. With the best will in the world, there are real dangers here of apparent bias. The World Economic Forum may appear to represent the interests of the wealthier segments of our global society and the custodians of both the ESI and EPI are located in two of the biggest and wealthiest universities in the wealthiest country on Earth, and for some it would seem to be a small jump from that background to a rather negative assumption that the indices have been designed to reflect well on the wealthiest. One could readily imagine calls to dismiss these pictures of the world as an index-centred version of 'fake news' or a case where 'facts are not facts'. I am sure that such claims would be met with fierce resistance by the custodians of the EPI and probably by those countries who do well in the EPI rankings. The following report from *Sage Magazine* (published by the School of Forestry & Environmental Studies, Yale University) describing the response of the Turkish government to its poor performance in the EPI speaks volumes:

> On January 27, a Turkish reporter for the Dogan News Agency wrote about Turkey's low ranking in the 2016 Yale Environmental Performance Index (EPI). Two days later, Dogan News published "Turkey on the Bottom," an exposé bemoaning Turkey's "retrogressive decade" of environmental performance that pushed the country's Climate & Energy and Biodiversity & Habitat index scores below those of Iraq, Syria, Libya and Haiti. Within hours, the article was shared thousands of times on Facebook. It drew a swift rebuke from Turkey's hardline government supporters and ignited a media firestorm that produced scores of articles.
>
> *(Mosteller, 2016)*

Ideally, of course, once would hope the Turkish government responded by actually doing something to improve its ranking, but the much easier and cheaper option is simply to respond by criticising the EPI as fake and lies. A sad reflection of the times we live in; it makes life so much easier if we only accept the indices that make us look good and reject all the others that do not as fake.

Conclusion

This chapter has explored the evolution of the ESI to the EPI and how that reflected a change in emphasis from sustainability to performance. Sustainable development is a complex subject that spans every other subject covered in this book. Trying to capture it with indices is a challenge and the approach often taken is to think in terms of a framework of indicators that could and often do include many of those provided in this book. Indeed, the United Nations has developed and released (in September 2015) a set of 17 Sustainable Development Goals (SDGs)[2] that, at the time of writing, have 169 targets and 304 indicators associated with them; what the UN calls the "global development framework". Countries have to report back on progress towards the targets by using the indicator framework, and the intention is to have achieved the targets by 2030. Whether there are too many or indeed too few targets and indicators, is a point of hot debate in the sustainable development community, especially as collecting the required data for many indicators may require significant additional cost, but the SDGs have been widely welcomed and we will return to them in Chapter 10.

At first glance, the EPI and its predecessor the ESI provide us with a different approach to that of the SDGs with its target and indicator framework. The EPI would seem to have much more in common with the HDI, a headline index presented as a league table designed to provoke improvement. Unlike the bewildering array of SDG indicators broken down by country we have just one, although such a degree of simplification does come at a cost as it has to be built on a foundation of many assumptions, all of which can be, and many have been, challenged. Also, unlike the SDG framework of indicators, which are primarily intended as instruments for governments to use to assess their progress towards their targets, the EPI, in common with many of the others in this book, although not all, has a headline focus and is intended to attract attention from the media as well as politicians. It may perhaps be a rather simplistic analysis as the SDGs have also attracted a lot of media interest, but the headline indicators seem to be designed with the glare of the media very much in mind. It is hard to imagine any media outlet willing to regularly publish stories on the 304 SDG indicators, but the creators of the ESI and EPI with their emphasis on simplifying complexity to a single number and using that number to allow countries to compare themselves seem to have set out from the start to attract a wider attention. Whether the ESI and the newer EPI have succeeded in translating attention to impact is another matter, especially in the post-truth world we seem to live in, but if they do bring about improvements then that can only be welcomed.

Notes

1 The ESI and EPI, and the various reports in which they were published, can be accessed online via dedicated websites of the two universities involved – Yale and Columbia: https://epi.envirocenter.yale.edu/, and http://sedac.ciesin.columbia.edu/data/collection/esi/.
2 The Sustainable Development Goals and the associated targets and indicators can be accessed via the United Nations: https://www.un.org/sustainabledevelopment/sustainable-development-goals/.

References

Bell, S and Morse, S (1999). *Sustainability Indicators. Measuring the Immeasurable*. Earthscan, London.
Bell, S and Morse, S (2003). *Measuring Sustainability. Learning by Doing*. Earthscan, London.
Ecologist and Friends of the Earth (2001). Keeping score: Which countries are the most sustainable? *Ecologist* 31 (3), 44.
Mosteller, D (2016). "Yale's lies?" EPI's rankings ignite national controversy in Turkey. *Sage Magazine*, 9 March 2016. Link: http://www.sagemagazine.org/turkey-2016-epi/.

New Scientist (2005). List of top green countries revealed. *Upfront*, 2 February 2005, p. 6. Link: https://www.newscientist.com/article/mg18524853-300-list-of-top-green-countries-revealed/.

World Commission for Environment and Development (WCED) (1987). *Our Common Future*, Oxford University Press, Oxford.

Further reading

Bell, S and Morse, S (eds.) (2018). *Routledge Handbook of Sustainability Indicators and Indices*. Routledge, London.

Dodds, F, Donoghue, D and Roesch, J L (2016). *Negotiating the Sustainable Development Goals: A Transformational Agenda for an Insecure World*. Earthscan, London.

6

POVERTY, INEQUALITY AND VULNERABILITY INDICES

Introduction

Many of us have an abject fear of poverty. The famous American comedian Julius Henry 'Groucho' Marx, he of Marx Brothers fame and, by the terms of his day, a very wealthy man indeed, once said:

> I always had a real fear of poverty. It came from years of living in boarding houses, bad hotels, bum clothes and cheap shoes.
>
> *(Taken from an article in the* San Francisco Examiner, *San Francisco, California, November 14, 1971, p. 182)*

Marx made his living from laughter, but when it came to poverty he was deeply serious. Because his family were so poor he had to leave school at the age of 12 and work. The Marx Brothers released many of their movies during the period of the Great Depression in the 1930s, although it is often said that the movie business was one of the few that were relatively untouched by the crash of the economy; during dark times people need an outlet. But Groucho's comment above refers to his early years during the late 1800s growing up in New York, and he is certainly not the only person to have developed a fear of poverty. Whether famous or not, none of us want to live in poverty, but for many alive today across the globe it is a reality.

We often think of poverty as being restricted to what we call the developing world, comprising distant places such as Africa, parts of Asia, and Latin America, but as the Groucho Marx comment above reminds us, poverty is often to be found on our own doorstep. Much can depend on what we mean by poverty. How do we recognise it? At first glance this may sound like a very silly question. Surely we all know what poverty looks like and, as Groucho noted, we can feel the symptoms if we ever find ourselves in that situation. But in practice it is not as simple as it may seem, and poverty may be perceived in comparative terms rather than being absolute. In a relatively wealthy country, poverty may indeed be regarded as having to wear cheap clothes and living in bad hotels, as Groucho mentioned, while in other parts of the world people may not even be able to afford even the cheapest of clothes or the worst hotel. Much can depend on context. But for those whose job it is to record these things for official statistics, it is important to be able to define what is meant by poverty before they are able to measure it. After all, following on from the "If you can't measure it, you can't improve it" sentiment expressed in Chapter 1, albeit with reservations, you certainly cannot measure something if you do not know what it is.

An important aspect of poverty alluded to in the title of this chapter is vulnerability. They go hand-in-hand, as those who are the poorest are often also the most vulnerable, and maybe that is part of the fear of being in poverty. We all have an innate sense of what being vulnerable means and can feel it in many contexts in our lives. From walking down a dark street late at night in a bad part of town to a sense of vulnerability during sickness, we can all appreciate what it means in a personal sense. But can we assess such vulnerability? There have been many attempts to do this, and in this chapter I will cover just one of them, the University of Notre Dame Global Adaptation Index (ND-GAIN),[1] which captures vulnerability to climate change at the level of the nation state. It is not included here because it is necessarily the best such index of vulnerability, but it does present a number of points of interest as to how vulnerability is conceptualised.

Poverty indices

Poverty has been defined in various ways over the years. Perhaps the simplest approach, and the one that still tends to dominate many people's perception, is to think in terms of a 'poverty line'; a level of income below which people are defined as being in poverty. The use of a poverty line in this way has a neat appeal. It is easily understood and the picture of a society presented by defining people into two camps – those who are poor and the rest – does serve to provide simplification for what is in reality a complex spectrum of categories found in most societies. It allows politicians and others to focus on helping those below the poverty line, although this raises the question of what happens to those who are just above it.

If we accept the notion of a poverty line as a way of defining who is in poverty and who is not, then the next obvious question is what should the poverty line be? What monetary value should be set for it? As can be imagined, much depends on the setting of this value for the poverty line. If it is set very high then it would be possible to classify almost the entire population as being in poverty, and if set too low then we could have a situation where no one is classified as being poor. There are two fundamental ways of setting the value for the poverty line:

1. *International standard*. Here the poverty line may be set at some level that can be applied globally, and an example is the assumption of US $2 per day. The advantage of this approach is that it allows for international comparisons.
2. *National standard*. Here the poverty line is set to best suit the local situation within a country rather than a global 'one size fits all' standard. International comparisons are possible by converting the national poverty line to an international currency such as the US dollar, but it needs to be remembered that the value of the poverty line will vary across countries.

In both cases the poverty line can move up or down as conditions change. For the national standard approach we could, for example, set the poverty lines relative to the median income in a country, typically as a percentage of the median income. The median, rather than the mean, is chosen because incomes across a population are often highly skewed with relatively small proportions having a very high income. If the mean was used then the poverty line would be pulled-up by that small proportion of higher earners. The median, the middle point of a distribution, is more appropriate in this context.

The poverty line indicator that is often used is called the Headcount Ratio: The proportion of the population below the poverty line.[2] Figure 6.1 presents the picture of the world if we could see the two versions of the Headcount Ratio from orbit, with the international poverty line set at US $1.90 per day. Darker shading means a higher value for the Headcount Ratio, which means a higher percentage of the population below the defined poverty line. First it is worth noting the many gaps as shown by the white shading, and particularly where they are. The gaps cover the more developed parts of the world, at least in terms of some of the other indicators presented in this book, and an obvious question to ask is why there are no Headcount Ratio's for those places? Surely it can't be for the lack of the required data on income levels and distribution? The answer is partly that these countries tend to use other measures to assess poverty rather than the Headcount Ratio. One example is the Townsend Index of Deprivation (TID) developed in the late 1980s (Townsend et al., 1988) and regarded at the time to be an effective measure of deprivation (Morris and Carstairs, 1991). The TID has four components that its creators believe capture deprivation:

1. Unemployment (% of those aged 16 and over who are economically active)
2. No car ownership (% of all households)
3. No home ownership (% of all households)

(a) Headcount Ratio based on a global figure of $1.90 per day.

(b) Headcount Ratio based on a national poverty line.

FIGURE 6.1 The world as seen with two versions of the Headcount Ratio. Darker shading indicates higher values, countries in white have no value for the Headcount Ratio. (a) Headcount Ratio based on a global figure of $1.90 per day, (b) Headcount Ratio based on a national poverty line

Source: Own creation based on data from the World Development Indicators (http://datatopics.worldbank.org/world-development-indicators/).

Household Overcrowding

These variables are combined, using equal weighting, to form the TID; the higher the value of the index then the more deprived and disadvantaged an area is thought to be. There are other such indices of deprivation, often varying between countries or even regions within a country and each are claimed to have their own advantages and disadvantages. The UK, for example, uses something called the Index of Multiple Deprivation (or IMD),[3] which is another of those complex aggregations of various components.

Ignoring the absence of Headcount Ratio's for the more developed countries in Figure 6.1, the picture presented is that of greatest extent of poverty being in Africa and that picture is consistent for both measures of the poverty line. But while the broad pattern is consistent there are differences in detail. Some countries do have higher Headcount Ratio's when measured using the national poverty line compared to the international: Mexico and South Africa are stand-out examples but there are many others. A 'scatter plot' of the two sets of Headcount Ratio data, based on international and national poverty lines, is presented in Figure 6.2 and while it looks like there is a relationship between the two, it is not a clear one. Notice how many of the data points are bunched towards the left-hand side of the graph while the others are widely spread out, even if there is a suggestion of a relationship. The choice of poverty line does matter a great deal. But which one of the two versions is correct? Which one should we use? Unfortunately there is no simple answer to these questions. You could argue that the international standard is the best one as it avoids the problem of political fudging, whereby national poverty lines can be changed to suit a political

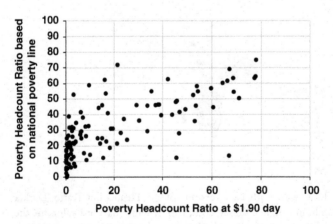

FIGURE 6.2 Relationship between the poverty Headcount Ratio calculated on the basis of $1.90/day (global) and national poverty lines. Dots represent individual countries

Source: Own creation based on data from the World Development Indicators (.http://datatopics.worldbank.org/world-development-indicators/).

agenda; in effect, a government could reduce the apparent degree of poverty in its country by lowering the poverty line. On the other hand, the one-size-fits-all approach imposed by the international standard may arguably be too blunt and it will ignore local conditions.

Besides the absence in Figure 6.2 of any comparable values of the Headcount Ratio for the more developed world, a further problem with the use of such a simple indicator is that it tells us nothing about the depth of poverty; how far below the poverty line people may be. It is theoretically possible to have identical Headcount Ratio's for two populations, but in one of these the people below the poverty line (set at, for example, US $2/day) may be only just below it (for example, an average of US $1.99/day) while in another they may be well below it (for example, an average of US $1/day). While the Headcount Ratio may be identical, there is a very big difference indeed between US $1.99/day and US $1/day: One is almost double the other. There are other indices of poverty that assess this gap but I will not cover them here. The reader is referred to Alkire et al. (2015) for more detail regarding the wide range of poverty indices.

Inequality indices

While income distribution is not a measure of poverty, it is important nonetheless. Indeed, in many countries there are increasing calls for differences between those who earn the most and those who earn the least to be made more transparent. One of the issues with the kind of national-scale indices I have discussed and reported in this book is that they tend to condense what can be very wide variation in an indicator within each country. Thus, for example, we end up with a single value of the Human Development Index for a country, which is great when creating league tables and maps to allow for cross-country comparisons. But given the three elements of the HDI, one of which is income proxied by GDP (or GNI) per capita, then surely we would expect wide variation in those components within each country? This point has been acknowledged in the case of the HDI, and some countries do now produce disaggregated values of the HDI for regions; and the UNDP has also started to produce an inequality-adjusted version of the HDI.

Taking income as an example, there are various ways of assessing income distribution, but perhaps the most commonly used is the Gini coefficient. Values of the Gini range between 0 and 1, with 0 being perfect equality (everyone earns the same) and 1 being the greatest inequality (one person earns everything while everyone else earns nothing). The theory behind the Gini coefficient[4] rests upon the measurement of deviation from complete equality in income, with higher values representing greater deviation from equality. In Figure 6.3, for example, there are 200 households, each earning one currency unit. Thus as we accumulate households across the horizontal axis – from 1 household at the left-hand side to 200 households at the right-hand side – the accumulated income from that population goes from 1 to 200 units. The line in Figure 6.3b represents

(a)

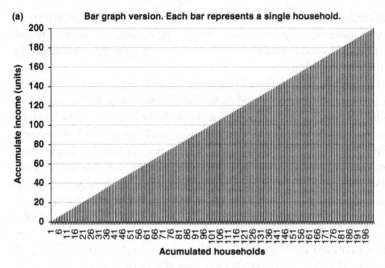

(b)

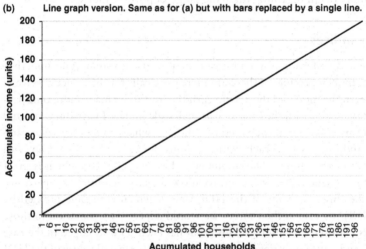

FIGURE 6.3 Accumulation of income for 200 households with each household having one unit of income. (a) Bar graph version, each bar represents a single household, (b) line graph version, each line represents a single household

complete equality in income between households and for this population there is no deviation at all from that line; hence the Gini coefficient is zero. In reality we do not see this for any population, even in countries that have embraced communism such as China, Cuba and indeed the old Soviet Union before its dissolution. In reality, for income distribution within countries we typically see something more like Figure 6.4. In this graph we still have 200 households and the total income for the country is still 200 units but we no longer have each household earning 1 unit. The poorest households are towards the left-hand side

(a)

Bar graph version

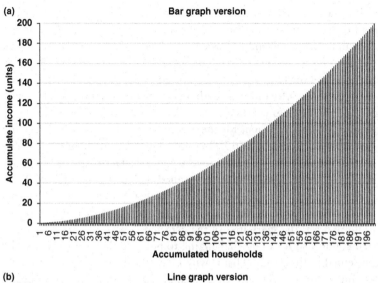

(b)

Line graph version

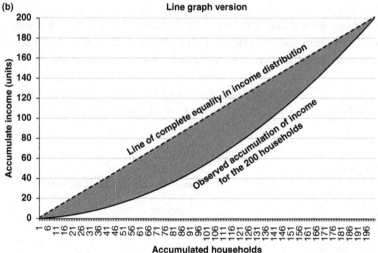

FIGURE 6.4 Accumulation of income for 200 households but with households differing in their income. Total money earned by the whole population is still 200 units, but it is not distributed evenly. The lowest-earning household is on the left-hand side, followed by the next lowest and so on until the highest-earning household is added towards the right-hand side of the graph. (a) Bar graph version, (b) line graph version

of the graph and the poorest of them all earns just 0.1 units. As we move from left to right along the horizontal axis, each household, starting with the poorest, is added until eventually we get to the richest household added towards the right-hand side of the graph. In this case, the richest household earns just under 1.9 units. When accumulated the total income for all households is still 200 units but now we have a far less equal distribution, a point stressed in Figure 6.4b where the line of complete equality is added for reference. The gap between the observed distribution and the line of complete equality, which is shaded in Figure 6.4b, can be measured and expressed as a proportion (0 to 1) or percentage (0 to 100%). In this case it comes to 0.3 (or 30%).

Country values for the Gini coefficient reported by the World Bank from 1980 to 2014 range between 0.16 and 0.66 with an average of around 0.4. However, for a single country the Gini coefficient can change over time. Figure 6.5 shows the change in the Gini coefficient for three countries – the US, Venezuela and Germany – between 1986 and 2013. The US showed a small but noticeable increase in Gini from below 0.4 to above 0.4, while Germany had a steady decline from 2006 to 2011, a period that spanned the economic crash of 2008. The Gini for Venezuela over that period was higher than for the other two countries, fluctuating between 0.42 and 0.53. But perhaps what is most noticeable from Figure 6.5 is the sparseness of the data across years. There are gaps in data for each country so it can be challenging to pull together a story across the period represented in the graph. That is why I have used bars here rather than a line, as we cannot assume what the Gini would be in the years that are missing from this dataset. This is one of the issues with the Gini coefficient, although it applies to all indicators and indices; the availability of good quality data, such as distribution of income between individuals or households, from which to calculate the index.

The picture of the world through the lens of the Gini coefficient is shown as Figure 6.6. Because of the paucity of data for all countries in the same year,

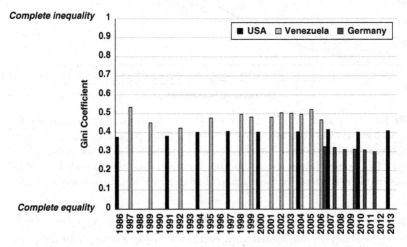

FIGURE 6.5   Changes in the Gini coefficient for the US, Venezuela and Germany from 1986 to 2013

Source: Own creation based on data from the World Development Indicators (http://datatopics.worldbank.org/world-development-indicators/).

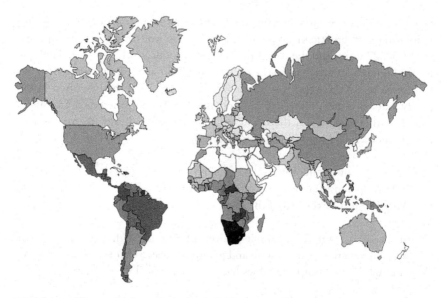

FIGURE 6.6 The world through the lens of the Gini coefficient. Darker shades represent higher values for the Gini coefficient and greater inequality. Countries in white have no data

Source: Own creation based on data from the World Development Indicators (http://datatopics.worldbank.org/world-development-indicators/).

the map is something of a fudge. The figures for the Gini are actually the most recent ones available for each country up until the latest year (2014) reported in the World Bank dataset. Hence there are some noticeable data gaps, mostly for the Arab countries of North Africa and the Middle East. Nonetheless, even from this patchy picture of the globe, the story is broadly one of some countries being more uneven than others. Sub-Saharan Africa and South America are particularly unequal, while some countries in Europe, particularly in the north such as Germany, are some of the most equal.

The Gini Index has a logic that can be displayed visually as shown in Figure 6.4 and it is also a commonly reported measure of inequality, with tables available via the websites of international agencies such as the World Bank. However, it does need to be said here that the Gini Index is by no means the only measure of inequality. Another one that is often used is called the Atkinson Index of inequality after its creator, the British economist Sir Anthony Barnes Atkinson (1944 to 2017) who specialised in researching income distributions. The Atkinson Index is provided by the following equation:

$$Atkinson\ Index = 1 - \frac{Geometric\ mean\ income\ of\ population}{Arithmetic\ mean\ income\ of\ population}$$

In our example above for the Gini Index, the unequal population of 200 house-holds had an arithmetic mean income of 0.9999775 units and geometric mean of 0.820532527 units, and the Atkinson Index works out as 0.179449011. Apologies for the long line of numbers to the right of the decimal point; we normally avoid this with data of this type as it implies a level of precision that we often cannot justify, but with this theoretical example the margins are fine ones. The higher the index then the greater the inequality. The Atkinson Index has a number of key advantages over the Gini Index, most notably:

1. Simplicity of calculation
2. Subgroup consistency
3. Weighting at lower end of the scale

As can be seen from the example above, all that is needed to calculate the Atkinson Index are the arithmetic and geometric mean incomes of the popula-tion, and while the Gini Index does have a visual appeal its calculation is not so straightforward.

The nature of the Atkinson Index allows us to estimate the index for sub-groups of the population so we can see where the relative inequality is greatest. This is not possible with the Gini Index. For example, with the population of 200 households used for Figure 6.3 we can divide them, starting with the lowest income, into subgroups of 10 households each, and calculate the Atkinson Index for each of these subgroups. The result is shown in Figure 6.7, with 20 values for the Atkinson Index calculated for the 20 subgroups, with the subgroups towards the left-hand side being the poorest and those on the right-hand side the richest. If inequality declines in any one subgroup of the population, perhaps between regions or genders for example, and inequality in all the other subgroups remains

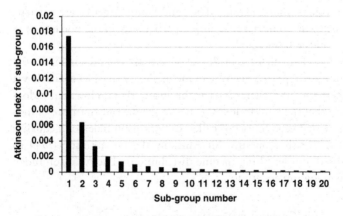

FIGURE 6.7   Atkinson Index for 20 subgroups of 10 individuals each based upon the same data used to calculate the Gini Index in Figure 6.4

the same, then overall the Atkinson Index will decline. This nuance can be very useful in terms of guiding interventions to help address inequality.

Finally, the Gini Index provides equal weight to all parts of the distribution, be they rich or poor. The Atkinson Index is different as it allocates more weight to the lower end of the income distribution, as can be seen in Figure 6.7. If you think of the overall inequality as being comprised of the bars in the graph then the biggest contributions to inequality are coming from the poorest groups. This helps focus minds in terms of policy and other interventions as it those groups where the needs are clearly the greatest.

The Atkinson and Gini indices of inequality can be applied to many variables, not only income. The Atkinson Index, in particular, has found its way into a number of indices covered in this book. In Chapter 7 we will meet its use within the Happy Planet Index,[5] one of my favourite indices, but it has also been used to create an inequality-adjusted version of the Human Development Index.

Vulnerability indices

Poverty and vulnerability are often the two sides of the same coin. Vulnerability, or its opposite, resilience, is usually expressed in terms of an ability to withstand stresses and shocks. Stress is a longer-term pressure within a system, such as, for example, a gradual increase in population density, while a shock is something more sudden, such as, for example, the economic crashes of 1929 and 2008, or an incidence of flooding or drought.

There are various indices of vulnerability that span environmental, social and economic stresses and shocks or indeed combinations of them. The example discussed here is the University of Notre Dame Global Adaptation Index, or ND-GAIN, and has been chosen because it spans environmental, social and economic stresses and shocks, and is primarily focussed on climate change. At the time of writing this book, the last is undoubtedly the biggest challenge facing humanity, and while I have covered the carbon and ecological footprints in Chapter 4, I did so from the perspective of carbon emissions (carbon footprint) and absorption (one of the components of the ecological footprint). But we also need to consider the vulnerability of countries with regard to the effects of climate change, and ND-GAIN assesses this in terms of three components:

1. Exposure to stresses caused by climate change
2. Sensitivity to climate change
3. Adaptive capacity to respond to the stresses brought about by climate change

The first of these assesses the exposure of a country to stresses (e.g. sea-level rise, changes in rainfall) brought about by climate change, while the second is the sensitivity to that change. These are quite different. A country may be highly exposed to stresses brought about by climate change but may not necessarily be sensitive to those stresses. The third element in the list is ability to respond and adapt to the stresses. Putting these three together, a country will be very vulnerable if it is highly exposed to stresses, highly sensitive to them and poorly equipped to respond.

ND-GAIN focuses on what its creators call "six life-supporting sectors" within each country: Food, water, health, ecosystem services, human habitat and infrastructure.[6] Each of these sectors is, in turn, represented by six indicators that span exposure, sensitivity, and ability to respond and adapt to climate change. The results as shown in Table 6.1 is a matrix of 36 (6 × 6) indicators in total to span all these life-supporting sectors.

In addition to these 36 indicators, which are meant to capture vulnerability of each country to stresses from climate change, ND-GAIN also includes indicators that span what the index creators call 'readiness'. This is different from adaptive capacity and is defined as:

> Readiness to make effective use of investments for adaptation actions thanks to a safe and efficient business environment.
>
> *(https://gain.nd.edu/)*

In effect, it is the ability of a country to leverage investment to help support the adaptive capacity.

The details of the ND-GAIN methodology do not need to be covered here, and indeed they have some similarity with those of the HDI, albeit with 36 indicators in the case of vulnerability rather than the 3 of the HDI, but it is interesting to note that the creators have opted for arithmetical means of the 6 indicators in each sector followed by the arithmetical mean across sectors. In the case of the HDI we have already seen how the United National Development Programme (UNDP) have in recent years opted for the geometric mean to avoid a compensation effect between components, but in the case of the ND-GAIN

TABLE 6.1 The ND-GAIN life-supporting sectors and their vulnerability to climate change

	Aspects of vulnerability to climate change			
Life-supporting sectors	*Exposure*	*Sensitivity*	*Adaptive capacity*	*Total*
Food	2	2	2	6
Water	2	2	2	6
Health	2	2	2	6
Ecosystem services	2	2	2	6
Human habitat	2	2	2	6
Infrastructure	2	2	2	6
Total	**12**	**12**	**12**	**36**

Source: Own creation based on information from The University of Notre Dame Global Adaptation Index (ND-GAIN) available at: https://gain.nd.edu/.

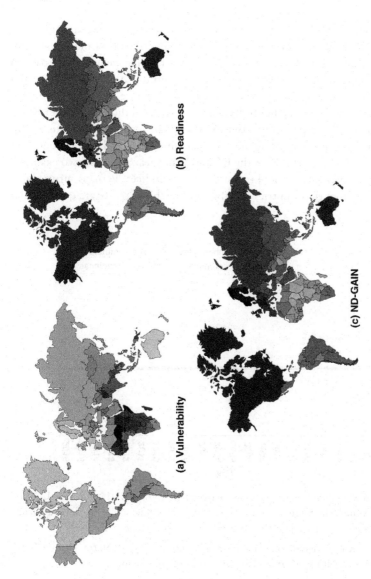

FIGURE 6.8 Vulnerability of countries to climate change and readiness to respond. Darker shades equate to higher values (greater vulnerability, better readiness, higher values for ND-GAIN)

Source: Own creation based on data from The University of Notre Dame Global Adaptation Index (ND-GAIN) available at: https://gain.nd.edu/.

(a) Vulnerability

(b) Readiness

(c) ND-GAIN

this approach was not adopted. Separate calculations are made for vulnerability and readiness and, in both cases, the index is adjusted to fall into a range of 0 to 1. Higher scores equate to greater vulnerability, which is bad of course, while for readiness higher scores mean better readiness (ability to leverage funds to support adaptive capacity), which is good. The two are combined as follows:

$$\text{ND - GAIN} = \left(\text{readiness score - vulnerability score} + 1\right) \times 50$$

This equation has the effect of scaling ND-GAIN from 0 to 100, with higher values being better (less vulnerability combined with better readiness) than lower values.

What does the world look like in terms of readiness and vulnerability?

Maps of the world set out in terms of vulnerability, readiness and ND–GAIN are presented in Figure 6.8. The story follows a familiar geographical pattern seen throughout this book. Africa comes out as having a high vulnerability and low readiness, while the developed world is the opposite. It would seem that poorer parts of the globe do indeed have greater vulnerability and a lower readiness to respond to climate change. But these maps are a snapshot in time, in this case for 2016. It is possible to see how vulnerability and readiness change over time and Figure 6.9 shows this for the US and Venezuela; the same two examples adopted in Chapter 2 when looking at economic indices. Perhaps surprisingly, given how different the countries are, their vulnerability scores are quite similar

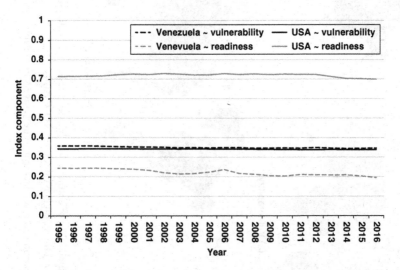

FIGURE 6.9 Vulnerability and readiness between 1995 and 2016 for the US and Venezuela. Higher values indicate greater vulnerability to climate change and greater readiness to respond to climate change respectively

Source: Own creation based on data from The University of Notre Dame Global Adaptation Index (ND-GAIN) available at: https://gain.nd.edu/.

and have been so from 1995 through to 2016. The big difference between them is in terms of the readiness component, where the US has values more than three times those for Venezuela. It is because of that difference in readiness that we see higher values of ND-GAIN for the US compared to Venezuela.

Conclusion

There are multiple indicators for poverty, inequality and vulnerability. The Headcount Ratio is probably the simplest and most widely reported indicator of poverty, as it is simply the proportion of a population below a nominal poverty line. The challenge is defining the poverty line. The value we allocate will depend in part on whether we opt to use a national or international value, and then, of course, it depends on what that value is. There are other poverty indices that go further than the Headcount Ratio, for example by looking at the depth of poverty (how far below the poverty line people may be), and yet others that take a different approach and assess deprivation.

Inequality is an important dimension to consider, especially as the tendency with the sort of nation-state level of indicators covered in this book. Clearly the notion of having a single value of an index covering a whole country is simplistic in the extreme, given the diversity that exists within each country. Yet few indicators have incorporated inequality, and the only one in this book that does is the Happy Planet Index. The issue of having one-index-fits-all has long been recognised by index makers and the response has typically been to develop versions of the index for geographical regions within each country rather than have an inequality-adjusted index.

Vulnerability is often expressed in terms of an ability to withstand stresses and shocks. Various indices have been developed to capture vulnerability and in this chapter only one of them, ND-GAIN, has been covered. ND-GAIN focusses on vulnerability to climate change and breaks it down into exposure, sensitivity and adaptive capacity. The worst-case scenario of high vulnerability to climate change occurs when a country is highly exposed to climate change, very sensitive to its impacts, and has little, if any, ability to adapt.

With all these sets of indices spanning poverty, inequality and vulnerability, we see a broadly similar pattern across the globe. The poorest parts of the planet also tend to be the most unequal and have the greatest vulnerability. But where is the cause–effect relationship? Does poverty lead to vulnerability and is inequality a result of poverty? It can be challenging to disentangle all these in simple terms of cause and effect, as we will explore further in Chapter 9.

Notes

1 The University of Notre Dame Global Adaptation Index (ND-GAIN) data and reports are available at: https://gain.nd.edu/.
2 The Headcount Ratio and data used in this chapter are available from the World Bank: https://data.worldbank.org/indicator.

3 Details for the Index of Multiple Deprivation (IMD) employed in the UK can be found at the Social Value Portal: https://socialvalueportal.com/indices-of-multiple-deprivation-in-the-uk/.
4 The Gini Coefficient data used in this chapter are available from the World Bank: https://data.worldbank.org/indicator.
5 The Happy Planet Index can be accessed via the following website: Happy Planet Index: http://happyplanetindex.org/.
6 (https://gain.nd.edu/).

References

Morris, R and Carstairs, V (1991). Which deprivation? A comparison of selected deprivation indexes. *Journal of Public Health Medicine* 13 (4), 318–326.
Townsend, P, Phillimore, P and Beattie, A (1988). *Health and Deprivation: Inequality and the North*. Croom Helm, London.

Further reading

Alkire, S, Roche, J M, Ballon, P, Foster, J, Santos, M E and Seth, S (2015). *Multidimensional Poverty Measurement and Analysis*. Oxford University Press, Oxford.
Atkinson, AB (2015). *Inequality*. Harvard University Press, Cambridge, MA.
Haughton, J and Khandker, S R (2009). *Handbook on Poverty + Inequality*. World Bank, Washington, DC.
Johnston, D C (ed.) (2015). *Divided. The Perils of our Growing Inequality*. The New Press, New York.
Raworth, K (2017). *Doughnut Economics Seven Ways to Think Like a 21St-Century Economist*. Chelsea Green Publishing, White River Junction, VT.
Sachs, J (2011). *The End of Poverty: How We Can Make It Happen in Our Lifetime*. Penguin Books, London.
Stiglitz, J (2013). *The Price of Inequality: How Today's Divided Society Endangers Our Future*. Penguin, London.
Wilkinson, R and Pickett, K (2010). *The Spirit Level: Why Greater Equality Makes Societies Stronger*. Bloomsbury, London.

7

HAPPY PLANET INDEX

Introduction

The words of Louis Armstrong's famous song 'What a Wonderful World', released in 1967, ooze happiness, even if the word 'happy' does not actually appear. We hear about trees of green and red roses, and later in the song we hear about babies crying and rainbows. Maybe it is my mental picture of Louis singing the song, but I cannot help but feel a sense of joy and happiness whenever I hear it. But what is it that makes us happy? This might seem like an easy question, but have a go at coming up with some answers and your list will quickly grow. Indeed, the chances are that the more you think about this question, the more answers you come up with. It is also more than likely that, as you travel along life's roads, then your answers will change. It is also more than likely that what makes you happy will not necessarily be the same as what makes me happy. So, given the very subjective nature of happiness, it seems almost impossible if not rather ludicrous that someone would even try to measure it. But, believe it or not, they have – and there is even a 'science of happiness'! In this chapter we will explore two indices of 'happiness'. The first of them is called the Happiness Index (HI) and it has some resonance with an index we will meet in the next chapter called the Corruption Perception Index (CPI), in the sense that it is based on surveys designed to ascertain people's views. Many of the other indices in this book are based on data collected primarily by governments, often from surveys but also from mandatory returns such as tax payments, birth and death records, employment figures, etc. For example, we have seen how government agencies can collect data on sales to create the GDP and its relatives, life expectancy data can be compiled from death certificates, schools and colleges keep records of the number of students, and so on. These are not data based on asking people about their experiences and feelings. But there are no such 'compulsory' equivalents that provide data for indicators of happiness. We have no choice but to ask people about this.

The second index makes use of the HI but extends it in two ways that are unique compared to all the other indices in this book. The Happy Planet Index (HPI)[1] is an attempt to generate a benefit:cost ratio, with benefit assessed in terms of 'happy life expectancy' and cost in terms of impact on the planet assessed as the Ecological Footprint, which we covered in Chapter 4. The HPI also attempts to incorporate inequality; one of the few 'headline' indices to do so, although there have been efforts in recent years to create what the UNDP call an 'Inequality-Adjusted' Human Development Index (IHDI).

A science of happiness …. Really?

Yes – really … Believe it or not, there really is such a thing as a science of happiness. Given that we all want to be happy, then understanding what is meant by happiness and how best to measure it does make sense. Happiness research has overlaps with research into well-being, and has some of its roots within psychology. Perhaps unsurprisingly, we would expect this to be a popular subject, there are many books and papers on happiness and in particular how to be happy. For the science of happiness, I can especially recommend the books written by Stefan Klein (2006) and Daniel Haybron (2013). A reviewer in *Der Spiegel* magazine has even written of Klein's book:

> "When you've finished reading this book, the inside of your head will look different," promises Klein. And he's right.

Few books can be said to change the inside of our heads!

But as a first step it is important to have a definition of what is meant by happiness, especially given the subjectivity noted in the introduction. Perhaps we can think of it as a certain state of mind, but we have many states of mind so how do we know which one relates to being happy? Most of us would probably agree that happiness is certainly a relative concept in the sense that what makes one person happy is not necessarily the same that would make another happy. In dictionaries happiness is often defined as the 'state of being happy', while being 'happy' is 'feeling or showing pleasure or contentment' (*Oxford Dictionary*). These do not seem to take us very far forward in terms of pinning the term down, but maybe that is to be expected. Indeed, the definition of happy seems to have significant overlaps with 'well-being', defined in the same dictionary as 'the state of being comfortable, healthy, or happy'. As happiness is also the state of being happy then it would seem that happiness is the same as well-being; a convenient tautology although it does seem an uncomfortable one. The discrepancy hinges on the inclusion of "or" in the *Oxford Dictionary* definition of well-being. Can one have a good sense of being comfortable as well as healthy but not be happy at the same time? These may not be so intricately linked as first imagined.

We could easily imagine how one aspect of our life, work for example, may be going well but other aspects, relationships perhaps, may not be. Nonetheless, and despite my doubts, in this chapter I will adopt the *Oxford Dictionary* approach of regarding the two terms – well-being and happiness – as being more or less interchangeable, but it is still not all that helpful in providing a means by which either can be measured.

Some organisations have proposed definitions of well-being that begin to provide a basis for assessment. One of them is the Organisation for Economic Co-operation and Development (OECD) in its publication entitled *Guidelines on Measuring Subjective Well-Being* (2013, p. 10):

> Good mental states, including all of the various evaluations, positive and negative, that people make of their lives and the affective reactions of people to their experiences … This definition of subjective well-being hence encompasses three elements:
>
> 1. Life evaluation—a reflective assessment on a person's life or some specific aspect of it.
> 2. Affect—a person's feelings or emotional states, typically measured with reference to a particular point in time.
> 3. Eudaimonia—a sense of meaning and purpose in life, or good psychological functioning.

The OECD definition may still seem rather vague, but it does provide three elements that form the basis for well-being. I do not intend to go into the history of measuring happiness here – there are some excellent texts that already do this, and the interested reader can find examples in the 'Further reading' at the end of the chapter. Instead, in the next section I will focus on how it is achieved for the Happiness Index and some of the challenges involved.

Happiness Index

The year 2012 saw the publication of the first World Happiness Report.[2] Of all the reports mentioned in this book as the vehicles for presenting indices to the world, this one has arguably the best title of them all. This first World Happiness Report was designed to coincide with a United Nations High Level Meeting called 'Wellbeing and Happiness: Defining a New Economic Paradigm' held in the same year. After 2012, World Happiness Reports were published in 2013, 2015, 2016 and 2017. They included a Happiness Index for most countries, and the results published for the World Happiness Report in 2017 (Helliwell et al., 2017) are shown in Figure 7.1. Darker shades imply higher values for the Happiness Index, which in turn means that people are happier. Even from a quick glance at Figure 7.1 it is obvious that the happiest places in the world seem

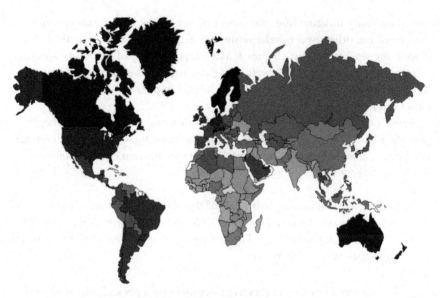

FIGURE 7.1   The Happiness Index for 2017. The darker the shade then the higher the values of the Happiness Index and the happier the country

Source: Own creation based on data available in the World Happiness Report for 2017 (http://worldhappiness.report/).

to be those that are the most economically developed, as we have seen in Chapter 2, and also those with the highest Human Development Index (Chapter 3). The unhappiest countries are mostly in Africa, it would seem, but the 'zone of unhappiness' also extends eastwards to India and China. While the map is a snapshot of happiness based on the 2017 report, the authors do point out that the results have been quite consistent between 2017 and the previous reports. But do these results suggest that money, or indeed human development, brings happiness? Surely this is far too simplistic? The question thus becomes how the Happiness Index assesses happiness?

The methodology set out for the Happiness Index in the 2017 report is summarised by the report's authors as:

> Our analysis of the levels, changes, and determinants of happiness among and within nations continues to be based chiefly on individual life evaluations, roughly 1,000 per year in each of more than 150 countries, as measured by answers to the Cantril ladder question: "Please imagine a ladder, with steps numbered from 0 at the bottom to 10 at the top. The top of the ladder represents the best possible life for you and the bottom of the ladder represents the worst possible life for you. On which step of the ladder would you say you personally feel you stand at this time?
>
> *(World Happiness Report, 2017, p. 9)*

It does not say it here, but respondents were 15 or over, and data were collected as part of the Gallup World Poll along with other sources. Gallup is an American research company founded in 1935, and is probably most famous for its opinion polls in politics, but since 2005 it has also undertaken world polls of citizens in 160 countries on matters of interest.

The Cantril ladder (Figure 7.2) mentioned in the quotation is named after its creator – Albert Hadley Cantril, Jr, a Princeton University psychologist, who lived from 1906 to 1969. The idea behind it is set out in his book published in 1965 entitled *The Pattern of Human Concerns* (Cantril, 1965). The question asked in Figure 7.2 could be thought of as 'life evaluation' mentioned in the OECD definition of well-being given earlier – a reflective assessment of a person's life – but obviously greatly simplified as the respondent is only being asked to provide a score from 0 (worst possible life) to 10 (best possible life). The scale in the ladder is referred to as 'self-anchoring' in the sense that each respondent has their own sense of what the 'best' and 'worst' possible lives are for them. Thus, the rung of the ladder they select is relative to those self-defining end points, and as a result of this internal relativity it does not necessarily mean that a rung selected by any one person is the same in absolute terms as that selected by another; one person's rung 5 is not necessarily the same as another person's rung 5. This may seem odd, but it is very difficult, if not impossible, to arrive at any objective sense of happiness given the obvious subjectivity involved here.

Another point, which is obvious but needs to be made nonetheless, is that our happiness can 'swing' up and down over time and indeed space. Our 'feeling' of happiness may change dramatically during a day and can also be influenced by

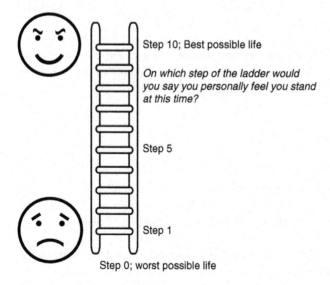

FIGURE 7.2   The Cantril ladder. The Happiness Index is the average of the steps selected per country based on asking this question in the years 2014–2016

146 Happy Planet Index

where we are or who we are with. Therefore, trying to assess something as volatile and subjective as happiness across a group of respondents and expecting that to represent the state of happiness in a single country does seem to be rather precarious.

Finally, it is worth making the point about scaling-up sample results to represent a country. The brief statement of the HI methodology noted above says that the samples upon which the HI is based is comprised of 'roughly' 1,000 respondents per year in each country. That might sound like a lot, but if a country has 60 million people then this represents a little less than 0.002% of the total population. Can we really use samples as small as this to represent the happiness of a whole country? It does seem very hard to believe that such samples can be used in this way to capture the views of a much larger population. But there are some well-established approaches that allow for this. For example, pollsters often use relatively small samples of the full population to make predictions of election results, although they have been seriously wrong many times, as they depend upon the sample being representative of the wider electorate as well as, of course, people voting as they claim they said once in the privacy of the voting booth. In reality, both of these can mean there are distortions. A proportion of the chosen sample may decline to provide an answer, and it is possible that those who decline may be more likely to answer in a certain way. In addition, people may respond one way when confronted by someone asking them questions in the street or on the phone, and in quite a different way when in private. Hence, there can be a significant gap between what people say and what they do. But at least in theory, it is possible to show that samples of a small percentage of a total population can be accurate representations, and much empirical evidence has been accumulated over many years that has helped inform methods designed to avoid the sort of issues noted above. Pollsters may still not achieve perfect predictions, but they are right a lot of the time.

There are mathematical equations that can help determine an appropriate sample size given the degree of accuracy we hope to get, and they do vary in terms of their assumptions and complexity. One of the simplest is provided in a statistics textbook written by Yamane and published in 1967:

$$S = \frac{N}{\left(1 + N e^2\right)}$$

In this equation the symbols are as follows:

S = appropriate sample size (the number we want to find)
N = total population
e = margin of error (not, in this case, Euler's number as used in natural logarithms). This can be thought of as the level of confidence we will have in the sample size. The lower the value then the greater the degree of confidence that the sample will be representative of the wider population.

By 'representative' we mean whether the sample accurately reflects the wider population in terms of characteristics such as age, gender, level of education, ethnic group, and so on. Clearly, if the wider population has an approximately 50/50 divide in terms of men and women, but we only include women in the sample then there is a danger that the results will not be scalable to the wider population. This sounds straightforward, but much depends upon the characteristics we deem to be relevant when scaling-up the results, and it is also quite possible that something important may be missed. Given that resources for carrying out surveys are almost always limited, then it is often the case that those selecting the respondents and asking the questions lean more towards convenience; for example, only including answers from those willing to provide them. If one group is reluctant to answer, they can become under-represented in the sample and this could have repercussions when extrapolating from the sample to the population.

If we assume the total population is 60 million and let us say we want a sampling error of around 5% (or 95% confidence in the representativeness of the sample) then the appropriate sample size using the Yamane (1967) equation turns out to be around 400. If we want a 1% sampling error (or 99% confidence in the representativeness of our sample) then the sample size becomes 'roughly 1,000'. The value of S using the Yamane equation will fluctuate with population size but not by much when we are talking of populations in the millions, which covers most countries, so the sample of "1,000 per year in each of more than 150 countries" used for the HI is about right. In reality, we might well end up having to contact many more people than this simply because, as noted above, not everyone contacted will respond, although this can depend on the means of making contact. Also, as so often in social science research, we are dealing with probabilities here and while we can be 95% or 99% confident about representativeness, there is still a chance that the sample may not be representative.

Finally, could it not also be the case that people asked questions set out along the lines of the Cantril ladder may give different responses depending on the way in which the question is asked? For example, with face-to-face interviews are people more inclined to say they are happy when compared to more impersonal methods such as the use of the internet or email? Are we uncomfortable with saying we are unhappy when speaking to a fellow human being compared with when we are asked to complete a form online? Could culture play a role whereby in some cultures baring one's soul like this may be seen as impolite or immodest? There are plenty of complicating factors involved with surveys but these are well-known and good planning can address many of them.

Assessing the wonderful world: Happy Planet Index

Given its title you might think that the Happy Planet Index is similar to the Happiness Index of the World Happiness Report, but it is quite different. Indeed, this is one of those indices that I often think has been inappropriately named, as 'Happy Planet' could give the unfortunate impression that it is the same as the

Happiness Index. In my view it should be called something like 'The cost of living a happy life', although admittedly this does not roll off the tongue and cost here is not monetary but cost in terms of our use of the planet's resources.

The HPI is produced by a non-governmental organisation called the New Economics Foundation (NEF) whose strapline is 'Economics as if the people and the planet mattered'. The NEF claims that the HPI:

> measures what matters: sustainable wellbeing for all. It tells us how well nations are doing at achieving long, happy, sustainable lives.
>
> *(happyplanetindex.org/about)*

The key word in this definition is sustainable and we have already met it at various points in the book, especially in Chapter 5. This is what distinguishes the HPI from the Happiness Index. The Happiness Index is a snapshot – it tells us how people are feeling at the time the surveys were undertaken, but nothing about how that happiness is to be maintained into the future or indeed what the cost is of achieving that happiness. As its creators point out:

> The Happy Planet Index gives us a clearer picture of how people's lives are going. It does this by measuring how long people live, how people are experiencing their lives directly, and by capturing the inequalities in those distributions instead of just relying on the averages.
>
> By also measuring how much natural resources countries use to achieve those outcomes, the Happy Planet Index shows where in the world wellbeing is being achieved sustainably.
>
> *(Jeffrey et al., 2016, p. 2)*

The inclusion of inequality and resource use are significant advances compared to the other headline indices present in this book. The headline version of the Human Development Index, for example, does not include inequality or resource use and has never done so since its inception in 1990. Critics have certainly raised concerns about the lack of these concerns in the HDI, but the UNDP has always been clear that they see the simplicity and consistency of the index as being paramount. In the Human Development Report of 1994 we find the following response to calls in the early 1990s for adding other dimensions to the HDI:

> The ideal would be to reflect all aspects of human experience. The lack of data imposes some limits on this, and more indicators could perhaps be added as the information becomes available. But more indicators would not necessarily be better. Some might overlap with existing indicators: infant mortality, for example, is already reflected in life expectancy. And adding more variables could confuse the picture and detract from the main trends.
>
> *(UNDP HDR, 1994, p. 91)*[3]

In fairness, the UNDP have created various versions of the HDI adjusted for different aspects, such as gender and inequality (for example, the Inequality Adjusted HDI), with the latter using the Atkinson Index discussed in Chapter 6, but the headline version of the HDI – the version that is highlighted in the Human Development Reports and the one that attracts most of the headlines – does not include such an adjustment.

The HPI is found for each country based on the following:

$$HPI = \frac{Inequality\ adjuted\ wellbeing \times inequality\ adjusted\ life\ expectancy}{Ecological\ Footprint}$$

Well-being equates to happiness as in the HI, and life expectancy is the same component as that included in the HDI. The inequality adjustment is achieved using the Atkinson Index, covered in Chapter 6, and for life expectancy the Atkinson Index for each country is found by:

$$Atkinson\ Index\ of\ Life\ Expectancy = 1 - \frac{Geometric\ mean\ life\ expectancy\ of\ population}{Arithmetic\ mean\ life\ expectancy\ of\ population}$$

Values of the Atkinson Index of Life Expectancy are known and published for countries by the UNDP, and these published values can be used to estimate the 'Inequality Adjusted' life expectancy:

$$Inequality\ adjusted\ life\ expectancy = \left(1 - Atkinson\ Index\ of\ Life\ Expectancy\right)$$
$$\times mean\ life\ expectancy$$

For well-being there are no published values for the Atkinson Index for countries, so the HPI creators had to estimate these.

$$Inequality\ adjusted\ wellbeing = \left(1 - Atkinson\ Index\ of\ Wellbeing\right)$$
$$\times mean\ wellbeing$$

As with the HDI covered in Chapter 3, the creators of the HPI make a few other adjustments to make the scales of the life expectancy and well-being components the same.

The product of the components in the numerator (the top part of the equation) can be thought of as a sort of happy life expectancy adjusted to account for the inequality of that happy life among the population. The inclusion of inequality is intended to avoid a situation where a small minority of a population can be very happy and long-lived and thus distort the value for the population as a whole.

The denominator (the lower part of the equation) is the price paid for the happy life described in the numerator. As we have seen in Chapter 4, the Ecological Footprint (EF) is a measure of the impact on the planet in terms of resource use. Therefore, the HPI can be thought of as a sort of benefit:cost ratio, and can be argued to be an advance on indices such as the HDI that do not include the cost of achieving life expectancy, income and education. A final scaling of the index is undertaken so that the HPI ranges from 0 to 100, with higher values representing a better return on happy life expectancy per cost to the environment.

As well as the inclusion of inequality and cost the reader will also be able to see from the above that the HPI does not have an economic component – there is no GDP/capita in the equation as there is with the HDI. This is not an accidental omission. As the NEF makes clear:

> People vote for political parties that they perceive to be most capable of delivering a strong economy, and policy makers prioritise policies that increase Gross Domestic Product (GDP) – the standard measure of economic growth above other goals. Doing so has led to short-termism, deteriorating social conditions, and paralysis in the face of climate change.
>
> *(Jeffrey et al., 2016, p. 2)*

and

> In fact, GDP growth on its own does not mean a better life for everyone, particularly in countries that are already wealthy. It does not reflect inequalities in material conditions between people in a country. It does not properly value the things that really matter to people like social relations, health, or how they spend their free time. And crucially, ever-more economic growth is incompatible with the planetary limits we are up against.
>
> *(Jeffrey et al., 2016, p. 2)*

This intention of countering a dominance of GDP in terms of policy and management, particularly with regard to the promotion of a more sustainable approach (Chapter 5), is shared by the creators of a number of indices, and sits in marked contrast with the alternative approach taken by those who created the Genuine Progress Index (covered in Chapter 2), for example. Indeed, we can identify a number of strategies taken by those attempting to create indices that counter the power of economic indices such as the GDP and set them out as a spectrum ranging from 'Fine-tune' to 'Avoid':

1. *Fine-tune*. Here the index creators work within the fundamental model of the GDP but seek to adjust it to accommodate hitherto uncosted positives and negatives. The GPI is an example of this approach.
2. *Dilute*. Here the GDP/capita is included as but one element in an index. This approach seeks to acknowledge the importance of economic income

but avoids its dominance by accepting that there are other important dimensions to development. The HDI is an example of this approach, with GDP/capita being one of three components.

3. *Avoid*. Here the GDP or indeed any economic indicator is not included at all. The HPI, EF and EPI are examples of this approach.

All these approaches are often included under a generic banner of 'Beyond GDP', promoted by Nobel Prize winners Joseph E. Stiglitz and Amartya Sen (also one of the influences behind the concept of human development covered in Chapter 3), amongst others (Stiglitz et al., 2010). Whether 'Beyond GDP' has diminished the dominance of GDP in the eyes of most policymakers, especially since the Great Recession of 2008, is open to question. I will return to this need to understand the influence of indicators and indices among those that are meant to be influenced by them in Chapter 10.

Is the planet happy and wonderful?

Figure 7.3 shows how the world looks through the lens of the HPI. Darker shades represent higher values for the HPI, although there are some countries, notably

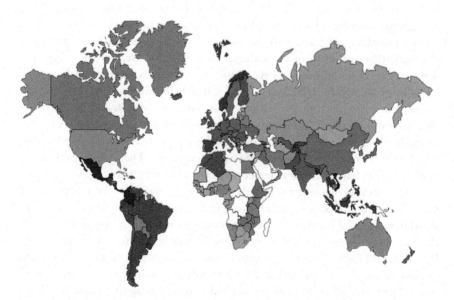

FIGURE 7.3   The world seen in terms of the Happy Planet Index for 2016. Darker shades equate to higher values of the HPI, while countries in white have no values for the index due to missing data

Source: Own creation using data from the Happy Planet Index website (http://happyplanetindex.org/).

in Africa, with missing data. The NEF make it clear that the missing data are not the NEF's fault:

> We rely on the availability of robust data from the United Nations, Gallup World Poll and the Global Footprint Network to calculate the Happy Planet Index score for each individual country. Unfortunately, that data isn't available for every country.
>
> *(happyplanetindex.org/about)*

A relatively complex index such as the HPI requiring data from a number of sources, each of which in turn is dependent upon data coming from other sources, may often be prone to these problems of missing data. We will return to the problem of missing data in the following chapter on the Corruption Perception Index and also in Chapter 10, when we explore the 'new index on the block' – the Sustainable Development Goals Index.

The HPI is relatively high in Latin America, Europe and parts of Asia and low in the countries of Africa that had values for the index as well as Russia, the US and Australia.

The HPI has a solid logic and certainly has had its adherents. It is one of my favourite indices precisely because it includes an adjustment for inequality, which few indicators do, and also because it has 'cost' in terms of resource use built in. It is the only headline index, by which I mean an index that forms the centrepiece or highlight of a report published by a major agency on a routine basis, which incorporates inequality and resource use. Variants of other indices are published that scale them in terms of inequality and/or resource use, but these typically appear as small sections in reports or in research papers rather than being the headline of the report. This does matter, as it is the headline that is often the key message that the press, policymakers, politicians, etc. absorb and use to frame their views and actions. These groups are unlikely to read research papers or read through what can be very lengthy and detailed reports to find sections where the headline index is adjusted for inequality or resource use. The creators of the HPI, to their credit, have built these concerns in from the very start.

But the HPI also has its detractors. At one level, this should not be a surprise as all the indices in this book have their critics and the HPI would not be expected to be any different. All of them are based on assumptions made by people and as a result these can be, and often are, questioned. One additional issue from Figure 7.3 is the lack of an HPI for a number of countries, especially in Africa. As noted above, this is often an issue with indices that require many datasets. Indeed, it is often not so much a matter of lack of data but lack of good quality, or what the NEF call 'robust', data although what is meant by quality here can be somewhat subjective. The other issue is the inclusion of well-being, which has echoes with the Happiness Index:

> How satisfied the residents of each country feel with life overall, on a scale from zero to ten, based on data collected as part of the Gallup World Poll.
>
> *(happyplanetindex.org/about)*

What was said above about the Happiness Index and its use of the Cantril ladder also applies to the HPI, of course, as well-being is a part of the index. Similarly, criticism has also been levelled at the Ecological Footprint, as we have seen in Chapter 4, and its inclusion with the HPI exposes it to the same criticism.

Conclusion

Measuring well-being (= happiness) with indicators and indices has attracted much attention and clearly does have a resonance with the wider population. Who would not want to be happy? But measuring the happiness of a country is prone to many methodological difficulties, not least the fact that it is subjective. Devices like the Cantril ladder allied with surveys can provide a snapshot, and their proponents can certainly argue that this is better than nothing, but the reader may still have doubts that happiness can really be captured by a single number and that results from a sample can really be extrapolated to a country. In this chapter I have made a case that survey results can be extrapolated with a degree of accuracy, although there are also mitigating factors that can work against this.

The Happy Planet Index (HPI) is, in my view, a commendable attempt to develop an index that includes well-being and life expectancy at its heart, but adjusts them in terms of inequality and relates this 'benefit' to the cost to the planet in terms of resource use. All of these are embedded in the index rather than included as 'add-ons'. There are few indices that have attempted to do this, although the HPI does have its issues. Nonetheless, the inclusion of a cost to the planet for benefits we obtain in terms of well-being and life expectancy is certainly a step in the right direction, as is the acknowledgement that the distribution of those benefits within society matters. As Louis Armstrong would probably have added in his song – a wonderful world has to be for all, not just a few.

Notes

1 The Happy Planet Index can be access via the following website: Happy Planet Index: http://happyplanetindex.org/.
2 The World Happiness Report can be accessed via the following website: http://worldhappiness.report/.
3 Further information can be found at Human Development Reports: http://www.hdr.undp.org/en.

References

Cantril, H (1965). *The Pattern of Human Concerns.* Rutgers University Press, New Brunswick, NJ.
Helliwell, J, Layard, R and Sachs, J (2017). *World Happiness Report 2017.* Sustainable Development Solutions Network, New York.
Jeffrey, K, Wheatley, H and Abdallah, S (2016). *The Happy Planet Index 2016. A Global Index of Sustainable Wellbeing. Briefing Note.* New Economics Foundation, London.

OECD (2013). *OECD Guidelines on Measuring Subjective Well-Being.* OECD Publishing, Paris.

Stiglitz, J E, Sen, A and Fitoussi, J-P (2010). *Mismeasuring Our Lives. Why GDP Doesn't Add Up.* The New Press, New York.

Yamane, T (1967). *Statistics: An Introductory Analysis.* 2nd edition. Harper and Row, New York.

Further reading

Haybron, D (2013). *Happiness. A Very Short Introduction.* Oxford University Press, Oxford.

Lehmann, S and Klein, S (2006). *The Science of Happiness: How Our Brains Make Us Happy – and What We Can Do to Get Happier.* Da Capo Press, Cambridge, MA.

8

CORRUPTION PERCEPTION INDEX

Introduction

It goes without saying that, for the most part, criminals want their crimes to be hidden. So why have I said it? Well, why the criminal may not want the rest of us to know that a crime has taken place, we certainly do. Indeed, while you may not feel like you have been the victim of a crime there can be repercussions for all of us from criminal activity. Insurance fraud, for example, drives up the size of premiums for all, including the innocent. There are many other crimes that carry a cost not only to the immediate victims but to wider society, and corruption is one of them. While corruption is often regarded as being an issue within the public sector, it is as well to remember that corruption can take place in all settings, including the private and voluntary sectors and may not just be based upon an exchange of money. Four definitions of corruption, three of them taken from dictionaries, are as follows:

> Dishonest or fraudulent conduct by those in power, typically involving bribery.
>
> *(Oxford Dictionary)[1]*

> Corruption is the abuse of entrusted power for private gain. It can be classified as grand, petty and political, depending on the amounts of money lost and the sector where it occurs.
>
> *(Transparency International, https://www.transparency.org/)*

> Corruption is dishonesty and illegal behaviour by people in positions of authority or power.
>
> *(Collins Dictionary)[2]*

Illegal, bad, or dishonest behaviour, especially by people in positions of power.

(Cambridge Dictionary)

In all four of these definitions we have the mention of abuse of power, but perhaps surprisingly only one of them, the one from a non-governmental organisation called Transparency International (TI), speaks of corruption being driven by a sense of private gain. The other three definitions refer to dishonest behaviour, and two of them to illegal behaviour, but they do not mention the drivers for this. Indeed, while 'illegal' is clear enough, 'dishonest' does sound like a rather vague and value-laden term. The *Oxford Dictionary* defines it as:

Behaving or prone to behave in an untrustworthy, deceitful, or insincere way.

The problem is that one person's sense of something being "untrustworthy, deceitful, or insincere" may not necessarily be the same as another's. Taken to an extreme, one could imagine someone in power acting in what they believe to be in a trustworthy and sincere way while others may not share that view. Indeed, the motivation behind the behaviour of the person in power may not necessarily be personal gain. Hence, they may be regarded as corrupt by some but they would probably not share that opinion. Definitions do matter but sometimes they may not help that much.

The TI definition is arguably the best of those given above, and itself echoes a much older and shorter definition of corruption given by the World Bank in its *World Development Report* of 1997, which is probably still the most reported definition (Liu, 2016):

Abuse of public power for private gain.

(World Bank, 1997, p. 102)

Such abuse of power for private gain does have a significant negative impact on the lives of many, but it also has a drag effect on development. Over many decades, scholars have written about the negative impacts of corruption, but I still find the following quotes, albeit some 15 to 20 years old, very apt (emphases are mine):

[Corruption] has distorted development priorities, led to massive human and financial capital flight, and undermined social and political stability ... corruption is deeply damaging to the *social and political fabric*, to investment, and to economic growth.

(Doig and McIvor, 1999, p. 660)

It is difficult to overstate the economic and *social significance* of corruption.

(Hisamatsu, 2003, p. 1)

I like these because they both go beyond a narrow vision of corruption being solely an economic issue; a matter of monetary loss. The first of them speaks of the damage that corruption does to the "social and political fabric" of a society while the second also includes the "social significance" of corruption. This is important as corruption can become embedded within a society and be regarded as the way of 'making things happen'; it becomes the norm, although few may feel that this is a good thing. The World Bank, in its seminal *World Development Report* of 1997, further illustrates this point

> Corruption violates the public trust and corrodes social capital. A small side payment for a government service may seem a minor offense, but it is not the only cost – corruption can have far-reaching externalities. Unchecked, the creeping accumulation of seemingly minor infractions can slowly erode political legitimacy to the point where even non-corrupt officials and members of the public see little point in playing by the rules.
>
> *(Wold Bank, 1997, p. 102)*

It is also necessary to add that "corruption, despite claims to the contrary, is not culture specific". (World Bank, 1997, p. 9). Corruption is not accepted in some cultures and rejected in others, and, as we will see later in this chapter, there are no countries in the world immune to corruption. Indeed, according to PricewaterhouseCoopers, a multinational service provision company:

> It is estimated that more than US $1 trillion is paid each year in bribes globally, and that US $2.6 trillion is lost to corruption. That's 5% of global GDP – and the true figure is probably even higher.
>
> *(www.pwc.com/gx/en/services/advisory/forensics/five-forces-that-will-reshape-the-global-landscape-of-anti-bribery-anti-corruption.html)*

And

> Corruption remains the single biggest issue facing today's society.
>
> *(www.pwc.com/gx/en/services/advisory/forensics/five-forces-that-will-reshape-the-global-landscape-of-anti-bribery-anti-corruption.html)*

But it is obvious that those who are corrupt want to hide that fact, except from their immediate victims of course. So how can we try and assess the level of corruption? Trying to ask those who are corrupt will get you nowhere. One could ask their victims, of course, but they may be afraid to speak for a number of reasons. If they identify someone who they have paid a bribe to then it may often come down to one word against another; proof may be hard to provide. Even worse, the corrupt may seek to punish the complainant and/or their friends and family. After all, and as noted in all the definitions of corruption given above, there is a strong disparity in power here and those who are the victims are often vulnerable. So, one may be left to try and somehow gauge the level of corruption indirectly.

Measuring something that is hidden from us is not a new challenge. Indeed, one of the most exciting developments in cosmology in recent years has been the idea that the vast bulk of the mass of the universe, perhaps 85% of all matter, is made up of something called dark matter. Dark matter does not shine when illuminated by starlight, and it does not interact very strongly with 'normal' matter; the stuff we are used to all around us. Hence, we cannot see it or feel it, but it seems that the universe has a lot of it, so how do we know it is there? Astronomers measure the extent of dark matter indirectly; by its gravitational influence, albeit low, on normal matter and the way in which it can bend the path of light through its gravity.

So, measuring the hidden can be done.

If astronomers can measure dark matter in the depths of space many thousands of light years from Earth then surely we can assess something much closer to hand and as familiar as corruption? The trick, as with dark matter, is to use an indirect approach, and there are various options. We could, for example, try to measure the spending of proceeds from corruption, perhaps by looking at the ownership of assets (large houses, cars, etc.) or construction projects in places where corrupt officials may live or are from. We could, for example, use high-resolution satellite imagery to detect whether there has been a surge in building work or asset ownership in places where those who are suspected of being corrupt may live. An alternative would be to look at spending behaviour and then asking people to account for the money they are spending: Where did it come from? However, the indirect approach taken by many has been to assess people's perception of corruption. Admittedly, perceptions can be deceptive and people can magnify or perhaps minimise their real experiences. They are often built from a combination of direct experience and, to be frank, hearsay derived from conversations with others who, equally, may or may not have been directly affected. Even asking apparently objective questions about how much was paid out as part of a corrupt exchange is open to distortion. Human beings are not always entirely reliable witnesses; we all too often create and change our memories of events. It is imperfect, but perception does provide a route and, after all, one could argue that, if the idea is to compare countries, then such imperfection is distributed evenly across the world.

This chapter will focus on one of the indices that has been designed to assess corruption at the level of the nation state; namely the Corruption Perception Index or CPI.[3]

The need for transparency

The CPI was created by a non-governmental organisation called Transparency International based in Berlin, Germany, but with many offices throughout the world. It was certainly not the first attempt to assess corruption at the level of the nation state, researchers have for many years undertaken academic surveys designed to explore corruption, but the CPI was the first attempt to routinely publish an index of corruption that would allow cross-country comparison. The motivation was to shine a light into the murky world of corruption. As TI say on their website:

> What does a number mean to you? Each year we score countries on how corrupt their public sectors are seen to be. Our Corruption Perceptions Index sends a powerful message and governments have been forced to take notice and act.
>
> Behind these numbers is the daily reality for people living in these countries. The index cannot capture the individual frustration of this reality, but it does capture the informed views of analysts, businesspeople and experts in countries around the world.
>
> *(www.transparency.org/research/cpi/overview)*

As with all the indices in this book, the idea of the CPI was to bring about a desired change, albeit 'desired' from the perspective of those creating the index who saw it as way of shining a light onto this murky world. Indeed, as with a number of the other indices, the use of country league tables was also adopted by TI for the CPI and the rationale is the same: To name and shame. Who would want to be labelled as the most corrupt country on the planet? It is hardly a title that inspires confidence in the electorate, and even the most dictatorial, undemocratic, thick-skinned and corrupt of politicians would find this a hard label to live with given that it is likely to reduce investment, international aid and damage relations with other countries.

Corruption Perception Index

One way of measuring the extent of corruption is to ask those who have been negatively impacted by it or, in other words, those that pay the cost. While this does make sense, given that it is highly unlikely that those who receive payments will admit to doing so, let alone respond honestly to survey questions, asking the 'payers' also has its problems. For example, it may be the case that the payers under- or overestimate the level of corruption they have experienced, perhaps because of embarrassment or perhaps because they are reluctant to admit engaging in an illegal practice. After all, you could argue that those paying the bribes are just as guilty as those who receive them, and this may act in the opposite direction and discourage people from reporting corruption. Also, if you believe a particular place to be corrupt then you will look harder for it; a self-reinforcement of an established bias. Nonetheless, there have been many surveys over the years to explore people's perceptions of corruption. Indeed, you might be surprised, as I was at first, by how many of these surveys exist. In itself this is testament to the importance of corruption. For example, the 2017 version of the CPI used the following sources, with the year of publication of the findings given in parentheses:

1. African Development Bank Country Policy and Institutional Assessment (2016)
2. Bertelsmann Stiftung Sustainable Governance Indicators (2017)
3. Bertelsmann Stiftung Transformation Index (2017–2018)
4. Economist Intelligence Unit Country Risk Service (2017)
5. Freedom House Nations in Transit (2017)
6. Global Insight Country Risk Ratings (2016)
7. IMD World Competitiveness Center World Competitiveness Yearbook Executive Opinion Survey (2017)
8. Political and Economic Risk Consultancy Asian Intelligence (2017)
9. The PRS Group International Country Risk Guide (2017)
10. World Bank Country Policy and Institutional Assessment (2017)
11. World Economic Forum Executive Opinion Survey (2017)
12. World Justice Project Rule of Law Index Expert Survey (2017–2018)
13. Varieties of Democracy (2017)

Each of the surveys attempts to assess countries based upon perceptions of corruption, and the respondents are typically those who visit many countries in the course of their work, such as business people. Some will have their own index of corruption and publish a league table ranking of nations based on that index. Methodologies also vary significantly between the surveys, including sample size, sample selection and the questions asked about corruption. Some have results for one year while others present results for several years. This diversity of surveys, each with its own method, can be rather bewildering and Transparency International came up with the idea of capturing all this information within a single index – the CPI.

The construction of the CPI has changed over time and the details need not be given here. The key point is that the CPI is an amalgamation of many individual surveys, all of which have one thing in common – they attempt to assess perception of corruption. One interesting feature of the CPI, which also applies to an extent to many of the indices covered in this book, for example the Happy Planet Index covered in Chapter 7, is how to address the problem of missing data. Even from a cursory glance at the list of surveys above for the CPI of 2017 it is obvious that some of them are regional rather than global in their coverage. Therefore, for some countries the data coverage will be incomplete. How does TI fill those gaps?

> Since many of the sources used for the CPI do not have a global coverage, the missing values for these sources are imputed for the baseline year.

Imputation is a statistical approach to filling gaps in data by making use of the data that does exist and making a reasoned estimate as to what the missing value could be. This is not about guessing what any missing values may be but using existing data to provide a rationale. It is by no means a perfect approach, as much depends as always on assumptions. The creators of the CPI for 2017, for example, used the following seven sources, all of which have a greater global reach in terms of their data collection to perform the imputation. These sources were selected because they covered more than 50% of the countries that were included in the CPI. The percentage coverage is provided in parentheses.

- Bertelsmann Foundation's Transformation Index (62%)
- Economist Intelligence Unit Country Risk Ratings (63%)
- Global Insights Country Risk Ratings (99%)
- Political Risk Services International Country Risk Guide (66%)
- World Economic Forum Executive Opinion Survey (64%)
- World Justice Project Rule of Law Index (55%)
- Varieties of Democracy Project (84%)

Imputation sounds a bit like magic. How can missing data be created, and how confident can we be that such created data are reliable? Other indices, even the HDI, have often faced similar issues with data gaps, and the approaches taken have varied. For example, we could simply acknowledge the gap and give the indices a null value, as with the Happy Planet Index in the previous chapter. Alternatively, we could simply make an estimate of the missing values by looking at corresponding values for peer countries and take an average of them for the country with the missing data. The latter approach has been taken by the creators of the Sustainable Development Goal Index; at the time of writing, this is a relatively new index that is beginning to gather a lot of traction in the realm of sustainable development. There are other approaches, such as the use of statistical techniques, that have been applied to fill gaps, and we will explore one of them – regression – in Chapter 9. Whatever approach is chosen, there will always be some pros and some cons associated with it.

The corruption scores from all of the 13 sources are rescaled so that they fall between 0 (highest perceived corruption) and 100 (lowest perceived corruption). The average is then taken of these rescaled scores to give the CPI. Lower values of the CPI equate to higher perceived levels of corruption and higher values equate to less perceived corruption. Note the emphasis here on *perceived* levels; it is critical to keep that in mind when trying to interpret the CPI.

Unlike many of the other indices covered in the book, the CPI does have the advantage of being calculated from multiple independent sources, all exploring the same issue from a variety of different directions, and this allows for the calculation of what is called the 'confidence interval' (CI), which sets out the bounds (minimum and maximum) where we would expect the mean to be. An example of the calculation of the CPI for 2017 is provided in Table 8.1 for the city state of Singapore. Of the 13 potential sources listed above, only 9 of them included

TABLE 8.1 Calculation of the Corruption Perception Index (CPI) for Singapore

Source	Score
African Development Bank CPIA	–
Bertelsmann Foundation Sustainable Governance Index	–
Bertelsmann Foundation Transformation Index	73
Economist Intelligence Unit Country Ratings	90
Freedom House Nations in Transit Ratings	–
Global Insight Country Risk Ratings	83
IMD World Competitiveness Yearbook	90
PERC Asia Risk Guide	92
PRS International Country Risk Guide	79
World Bank CPIA	–
World Economic Forum EOS	90
World Justice Project Rule of Law Index	85
Varieties of Democracy Project	77
Number of sources listed above	9
Average score (CPI)	**84**
Standard deviation of scores in the list above	6.78
Confidence interval for the average score (95%)	4
Upper confidence interval (average + CI)	88
Lower confidence interval (average – CI)	80

Note: These data are for the CPI 2017. Dashes indicate that no scores were available for Singapore from that source.

Source: Own creation based on CPI data accessed via the website of Transparency International (https://www.transparency.org/).

estimates for Singapore, and the table sets out the scores (altered to a scale of 0 to 100) achieved for each of the surveys. Please remember that higher values equate to less corruption and vice versa. The average of the 9 scores gives a value of 84, which is the CPI 2017 figure for Singapore. But given that we have 9 estimates of corruption we can go further and measure the variation between the scores. We do this by calculating what is called the 'standard deviation' (sd), which can be thought of as an average absolute deviation of values from the mean; it is not exactly that, but it may help to conceptualise the sd in that way. We have to use the absolute values of deviations (i.e. negative values have to be changed to positive ones) because the average deviation around a mean is always, of course, zero. The higher the standard deviation then the greater the variation around the mean. The confidence interval is derived from the standard deviation and an assumption as to the degree of confidence we wish to have about the location of the mean. If we opt for 95% confidence then for these data the CI becomes 4. This equates to an upper value for the mean of 84 + 4 = 88 and a lower value for the mean of 84 – 4 = 80. We can have 95% confidence that the true CPI (our measured value of 84 is but an estimate of this true value) is somewhere between

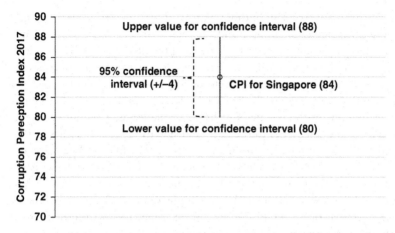

FIGURE 8.1   The mean score (equates to the CPI) and 95% confidence intervals for Singapore

80 and 88. We can represent this visually as shown in Figure 8.1, with the mean as the dot in the centre and the confidence interval as a bar that extends below and above the mean.

The use of confidence intervals is common in research and is a way of saying that we are not sure of the exact value of the true mean but we do have confidence that the true value exists within a range of values. The mean calculated in Table 8.1 is based on a sample of surveys (in this case nine), but there is variation between the surveys, so while we can calculate a mean from the sample (84 in our case for Singapore) there is some uncertainty as to what the true mean may be. Thus, we provide a range of values (the confidence interval) within which we can be reasonably confident that the true mean rests. While this is a transparent and honest way of representing uncertainty, rather oddly, at least in my view, this is often used as a criticism of research by people who are not themselves research-ers, as it can superficially imply a degree of indecision – a sense that the research-ers are not 100% sure of something. Thus, critics sometimes argue, if researchers are not sure where a value rests then that must discredit their case. This is a point that is often made, for example, with regard to global warming and the predic-tions of bodies such as the Intergovernmental Panel on Climate Change (IPCC). Their scenarios for future temperature change are based on complex models and often come with CIs either side of their predicted values, yet this is interpreted by some as discrediting the case for temperature increases as they claim that it shows the scientists are not 100% sure of their findings. But this is unfair as the researchers are being very open about what they are doing and very often there are uncertainties that exist around values such as means, and the CI is a way of estimating that.

With the CPI the inclusion of a CI is unusual among indices, and we have nothing like this for the Human Development Index or indeed any of the others presented in this book. With all of those indices all we are ever given is a single value of the index for each country and for each year. However, it is perhaps ironic that the CPI league tables (the headline presentations) are based solely upon the mean and do not include the CI. If a consumer of the CPI wishes to know what the CI is they need to search for it; just as I have had to do. The result of including the CI alongside the CPI can be seen in Figure 8.2. The graph is a plot of the CPI for the countries included in the league table for that year ranked from the highest values to the left-hand side to the lowest values to the right-hand side. The dots are the mean scores (the CPI) while the lines extending above and below each dot at a 95% CI. Note how some of the countries have very large CIs. For the Comoros the CI is 17, meaning that the CPI of 27 has a 95% CI band that extends from 12 to 42 – quite a range! Therefore, we have 95% confidence that the true CPI for Comoros is somewhere within that range but, while this is transparent, there is nonetheless a rather disconcerting degree of uncertainty here. Indeed, the CI is so large for some countries that it implies their league-table ranking could change by up to 10 places up or down. Even if the true CPI for Comoros is towards the upper end of the CI band it would still mean that the country sits towards the right-hand side of the distribution in Figure 8.2 and would still hardly be seen as good news. But even so, one could

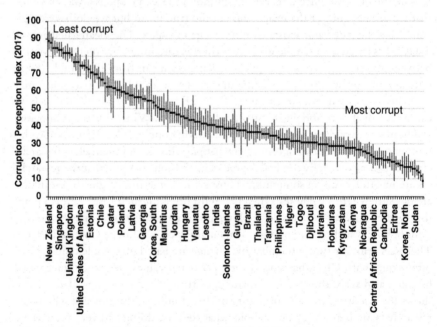

FIGURE 8.2   Country ranks based on the CPI for 2017. Most corrupt countries are to the right-hand side and least corrupt are to the left-hand side

Source: Own creation based on CPI 2017 accessed via the website of Transparency International (https://www.transparency.org/).

well imagine politicians in power making a lot of such a change and indeed the high degree of uncertainty! Therefore, to use a specific league-table rank for the CPI is arguably somewhat misleading.

Finally, as with all the other indices covered in this book, we have to repeat the usual warning that applying a single value of an index to a whole country can be highly misleading. The CPI is based upon the experience of those visiting or residing within a country, and it is quite possible that this experience is centred round commercial and urban centres rather than the country as a whole. It may also be the case that the experience is based on engagement with an extremely small segment, albeit a very powerful one, of the country's population. As it is based on perception, mostly of those who visit the countries on business, we cannot in any way claim that all the people in a country are corrupt or indeed not corrupt. The respondents who answered the surveys may have interacted with very few people, albeit those with a degree of power, and this is what has framed their perception. In Chapter 7 we covered the importance of using representative samples if we are trying to extrapolate to a wider population, but with the corruption surveys listed above the samples are based on the experiences of a group having contact with what is likely to be a small and highly unrepresentative subset of the population. As a result, the vast majority of a country's populace would not have been involved in the framing of the index that becomes associated with them. In that sense the CPI seems rather unfair, but it does at least capture some key relationships that will have an influence on decisions such as those surrounding investment. We also need to come back to what we said at the beginning of the chapter; that corruption, even amongst a few, can be "deeply damaging to the social and political fabric" of a country (Doig and McIvor, 1999, p. 660).

The world through the lens of the CPI

What does the world look like if we could visualise the CPI from orbit? Here I will take the CPI at face value and use the headline figures, the ones presented in the Transparency International league tables. Figure 8.3 presents the map of the world shaded according to the value of the CPI for 1995 (Figure 8.3a), the first one that Transparency International produced, and 2017 (Figure 8.3b); a difference of over 20 years. Darker shading in these maps equates to higher values of the CPI and lower perceived levels of corruption. Lighter shading is where corruption is perceived to be high. The first thing that may strike the reader is the paucity of coverage in the CPI 1995; there are many countries with no value at all. This does not mean that those countries have no perceived level of corruption; it just means that no data were available at that time. The map for 2017 has much better coverage. But, even with the partial picture for 1995 and the more complete picture for 2017, we can see a pattern. Perceived corruption tends to be higher for the southern and eastern hemispheres of the world, with the notable exception of Australia. A number of corruption 'hot spots' can be seen in Africa and southern Asia. The pictures also suggest a degree of consistency between

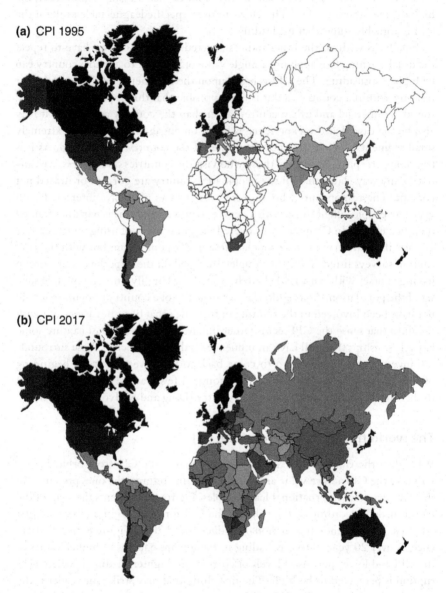

FIGURE 8.3 The world through the lens of the CPI. The darker the shading then the higher the value of the CPI (high values equate to lower levels of corruption). White shading means that no values were available for that year. (a) CPI 1995, (b) CPI 2017

Source: Own creation based on CPI data (1995 and 2017) accessed via the website of Transparency International (https://www.transparency.org/).

1995 and 2017, although it is not possible to be firm about this given the lack of coverage in 1995. Figure 8.4 is a plot of CPI 2017 against CPI 1995 and confirms the relationship between the two, even if the data are few in number and the scales are different, with the latter being due to a shift in methodology. Indeed, the two least corrupt countries in 1995, New Zealand and Denmark, are the same in 2017. Nine of the top 10 (least corrupt) countries in 1995 are also in the top 10 countries for 2017.

A consistency in the patterns of the CPI between 1995 and 2017 is also shown by the rankings of the countries in Figure 8.5. While the rank of any one country might change over that time, and indeed the methodology has also shifted, with final scoring of the CPI changing from values with a maximum of 10 to values with a maximum of 100, the patterns in both years show a steady decline with no 'jumps'. In 2017, the rate of decline in the CPI slows down towards the middle of the rankings before falling off more rapidly with lower values. But note how there is no country considered to be corruption free. There may be differences in perceived corruption but even the country at the top of the pile in 2017, New Zealand with a score of 89, is not completely free of the scourge.

What this picture tells us, among other things, is that the perceived level of corruption has remained relatively consistent over time. Is this because the underlying levels of corruption are more or less static or is it because perceptions become fixed and are hard to shift? One of the dilemmas with creating an index based on perception is that respondents may have framed their views over a long period of time and once a country gets a bad reputation it becomes

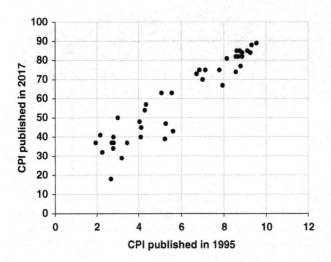

FIGURE 8.4   Relating the CPI 2017 to that of CPI 1995. Each dot represents a country

Source: Own creation based on CPI data (1995 and 2017) accessed via the website of Transparency International (https://www.transparency.org/).

(a)

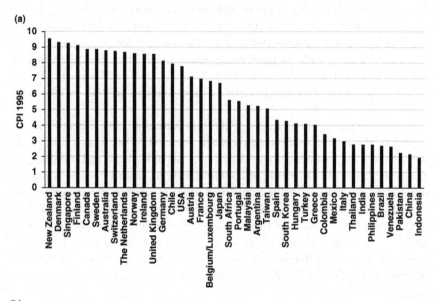

(b)

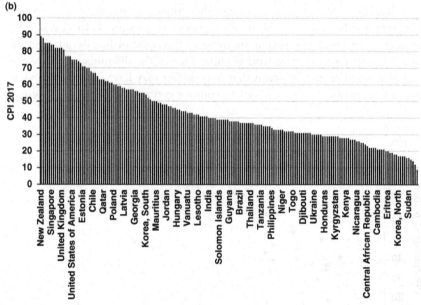

FIGURE 8.5 Country ranks for the first CPI of 1995 and that of 2017. The graph for the CPI of 2017 is the same as that of Figure 8.2 but without the confidence intervals. Countries have been ranked in order with highest values of the CPI (least corruption) to the left-hand side. (a) CPI for 1995, (b) CPI for 2017

Source: Own creation based on CPI data (1995 and 2017) accessed via the website of Transparency International (https://www.transparency.org/).

almost an expectation. It is hard to say one way or the other. One complicating factor in all this is that perceptions of corruption may not only be framed by personal experience but also potentially by the telling and retelling of the stories of others. Thus, it may be possible for a perception of corruption for a country to become embedded and hard to shift: Are we seeing some of that with the CPI?

Conclusion

This chapter has focussed on the CPI, an index based on what might seem to be a rather nebulous sense of perception arising from experience with corruption. Not only that, but the experiences that underpin the CPI are based on those from a rather narrow set of the global population; primarily business executives who regularly travel between countries. While on the surface there may be some similarity with the Happiness Index (HI) of the previous chapter they are quite different. The HI is derived from surveys conducted in a country, and, as we saw, one can calculate the appropriate sample size required in order to be able to have a degree of confidence that the sample would represent the views of a population. The sample is derived from people living in the country rather than from a more transient group of international travellers. Also, the questioning that forms the basis for the HI is arguably not sensitive in the same way that corruption is. Respondents may be unwilling to answer questions regarding their happiness, but being unhappy is not a crime, at least not yet! With the CPI not only is the value based on the views of a rather unusual group of people, but also their interaction with what is, arguably, another limited group (civil servants, politicians, other business people, etc.) in the countries that they visit. Thus care does need to be taken when interpreting the CPI; it should not imply that all the people in a country have a level of corruption set by the index. Also, given that the CPI has confidence intervals, then the exact placement of a country within a 'corruption hierarchy' is problematic to say the least.

However, for all its faults, the CPI does provide a very useful window on a subject with an importance that it is hard to overestimate, but one that lives in the shadows. The inclusion of a confidence interval for the CPI is a useful reminder that indices based on responses from surveys do have a degree of uncertainty. The same is true for the HI of course, although it is notable that CIs are not included in the headline publications of the Happy Planet Index. While the CIs for the CPI are also not included in the headline league table of that index, they can be found, albeit with a bit of searching.

The CPI tells us that there is no such thing, as yet, of a country free from corruption. The maps in Figure 8.3 do present a picture where corruption is much worse in some places than others, but the scourge is everywhere. Corruption is not something that is endemic in some places because of culture, but because of the presence of a large imbalance between those with power and those without.

Notes

1 The online version of the *Oxford Dictionary* can be accessed here: https://en.oxforddictionaries.com/.
2 The online version of the *Collins Dictionary* can be found here: https://www.collinsdictionary.com/.
3 The Corruption Perception Index data and technical notes can be readily accessed via the website of Transparency International and its various country offices: https://www.transparency.org/.

References

Doig, A and McIvor, S (1999). Corruption and its control in the developmental context: An analysis and selective review of the literature. *Third World Quarterly* 20 (3): 657–676.
Hisamatsu, Y (2003). Does foreign demand affect corruption? *Applied Economics Letters* 10 (1): 1–2.
Liu, X (2016). A literature review on the definition of corruption and factors affecting the risk of corruption. *Open Journal of Social Sciences* 4, 171–177.
World Bank (1997). *The State in a Changing World. World Development Report 1997.* Oxford University Press, Oxford.

Further reading

General texts on corruption, its importance and impact:

Cockcroft, L (2012). *Global Corruption: Money, Power and Ethics in the Modern World.* University of Pennsylvania Press, Philadelphia, PA.
Fisman, R and Golden, M A (2017). *Corruption: What Everyone Needs to Know.* Oxford University Press, Oxford, UK.

For a text on the challenges involved in the analysis of corruption the following is recommended:

Hough, D (2017). *Analysing Corruption.* Agenda Publishing, Newcastle-upon-Tyne, UK.

For those interested in the climate change debates and how scientific results are interpreted in different ways by people the following text are recommended:

Craven, G (2009). *What's the Worst That Could Happen? A Rationale Response to the Climate Change Debate.* Perigee, New York.
Dessler, A and Parson, E A (2010). *The Science and Politics of Global Climate Change: A Guide to the Debate.* 2nd edition. Cambridge University Press, Cambridge, UK.
Oreskes, N and Conway, E M (2010). *Merchants of Doubt. How a Handful of Scientists Obscured the Truth on Issues from Tobacco Smoke to Global Warming.* Bloomsbury Press, London.

9

SEEKING RELATIONSHIPS

Introduction

In this book I have presented a number of indices to the reader covering an array of issues common to us all across the globe; from economic performance, corruption, happiness, human development, poverty and inequality to the use of the Earth's resources and environmental impact. This is a wide-ranging set of issues, but it is important to stress that they are all interlinked. Thus, for example, poverty has a strong relationship with environmental degradation; people in poverty will prioritise their own survival even if this means that they degrade their environment. Corruption is interlinked with good governance in such a way that decisions made by those with power will not necessarily be in the best interests of the wider population. Hence, we would expect to see a decline in the state of human development with higher levels of corruption.

With the variety of indices covered in this book it is tempting to think about how they may be related to each other. After all, we have a variety of indices for each country and while there are issues to do with differences in the time periods over which data were collected for the indices, it is still tempting to see how they relate to each other and, if they do, then why that should that be. In this chapter I will do just that. I will look for relationships between the indices and provide some thoughts as to why any relationships may exist, or not. But first, of course, we have to define how we will identify a relationship, and in this chapter I have opted for a statistical approach. Not all readers will be familiar with statistical techniques so the chapter begins with a brief non-mathematical primer on correlation and regression, and this is followed by an exploration of the relationships between indices.

A brief statistical primer on relationships

We are often interested in looking for relationships between indices and there is a statistical technique that allows us to do this called the Pearson product-moment correlation coefficient (r). This longish name is often shortened to the Pearson correlation coefficient or just the correlation coefficient, as I have done here, and is named after Karl Pearson (1857 to 1936), an English mathematician who is regarded by many as one of the fathers of statistical analysis. But Pearson did not invent the idea of correlation. He expanded and developed the concept and mathematics of correlation discovered by another Victorian mathematician, Pearson's mentor in fact, Francis Galton. Once he had set out the core of the idea behind correlation, Galton sensed the wide applicability of the concept and went on to state in one of his papers:

> There seems to be a wide field for the application of these methods to social problems. To take a possible example of such problems, I would mention the relation between pauperism [poverty] and crime.
>
> *(Galton, 1890, p. 431)*

Galton thought that poverty could be a major causal factor for crime and indices at the level of the nation state can allow us to investigate this idea. I will indeed present a graph later in the chapter that links poverty with corruption, as suggested by Galton nearly 120 years ago!

It is not necessary here to go into the mathematical detail of the correlation coefficient, and some introductory texts on the subject are listed at the end of the chapter, but suffice it to say that it assesses the degree of linear correlation between two variables on a scale from −1 to +1. Values of +1 and −1 represent complete linear correlation between the two variables while a value of 0 means that there is no linear correlation. Note the emphasis here on 'linear'; the extent to which a relationship between two variables can be described by a straight line. The various graphs in Figure 9.1 provide visual illustrations of the correlation coefficient for countries using two indices, called A and B. In graph (a) the points are perfectly lined up in a straight line and the value of the correlation coefficient is 1. Here the value of Index B increases in line with the value of A. In graph (b) the points are again lined up perfectly but in this case the value of index B decreases as index A increases. In this case the value of the correlation coefficient is −1. Both graphs (a) and (b) in Figure 9.1 provide us with an extreme, where the values of one index is exactly in line with the other. In graph (c) of Figure 9.1 we have the other extreme, or at least close to it, as we have a random scattering of points with no apparent relationship between the values of the two indices, and the value of the correlation coefficient is almost 0.

While graphs (a), (b) and (c) in Figure 9.1 illustrate the extremes of the correlation coefficient (spanning the range from +1 to 0 to −1), there are situations that are more complicated and one such example is shown in graph (d). Here the

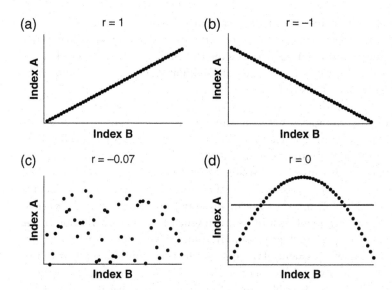

(a) r = 1

Index A

Index B

(b) r = −1

Index A

Index B

(c) r = −0.07

Index A

Index B

(d) r = 0

Index A

Index B

FIGURE 9.1   Examples of the correlation coefficient (*r*)

points for the two indices form a curve rather than a straight line and the correlation coefficient is 0, just as it almost is with graph (c). But something seems to be wrong here as we do seem to have an obvious relationship between the two indices in graph (d), albeit not a simple linear one as seen in graphs (a) and (b). It is important to remember that the correlation coefficient is a measure of the extent of a linear relationship between the two indices. Just because the correlation coefficient is 0, as in graph (d), it does not mean that there is no relationship at all between the indices; there may well be one but not linear in nature. Therefore, it is important to look at the scatter graphs between two indices rather than rely solely on the correlation coefficient to provide an indicator of linkage.

   In reality we are often faced with far messier plots between indices than we can see in Figure 9.1, and the examples I give later in the chapter will all fall into this messier category. The question then becomes: What value of *r* do we take as being statistically significant? What value for *r* gives us a degree of confidence that while the data are scattered there may be an underlying linear relationship between the two indices? Statisticians have worked this out and we can estimate cut-off points for *r* beyond which we can say that the statistic is statistically 'significant', and for the latter we typically employ the 5% (or 0.05 in decimal notation, the same as 1 in 20) cut-off. Thus, we can say that if the value of *r* is equal to or exceeding the 5% cut-off value then the chances of getting that value by chance are relatively low and it indicates that a linkage between the two indices is likely to exist. Because *r* can be negative as well as positive we are talking here of cut-off values that are both negative and positive. Hence, for example, while a value of *r* equal to −0.5 is technically less than say a 5% cut-off of −0.3 here I refer to it as exceeding (being greater than) the cut-off. Please also note that we are working with probabilities here and not certainties. We cannot say that a high value for *r* means that a linkage between the two indices exists beyond all doubt. All we can say is that it is highly likely. This might seem rather nebulous, but it is the way that many statistical tests work.

While the correlation coefficient provides us with a simple measure of linear relatedness between indices, with an easily understood scale of −1 to +1, it does have its disadvantages. The emphasis on linear relationship has already been mentioned, although in my experience it is often forgotten. It is also necessary to stress that just because two variables have a significant correlation coefficient, or indeed have a scatter of plots that suggest the two may be related in a cause–effect fashion, this does not in any way prove that they are. This may seem like an odd statement, as surely statistics such as the correlation coefficient are designed to provide such proof, and in some of the graphs of Figure 9.1 we do seem to have some evidence for a cause–effect relationship, but that is misleading. It is possible to produce statistically significant correlation coefficients between many pairs of variables, including indices, but that does not necessarily mean that they are directly related in terms of cause and effect. Indeed, there is an oft-used adage that correlation does not necessarily mean causation. It is possible that the variables may be indirectly related via a third one, but this may require further exploration. Hence, cause–effect interpretation of correlated variables needs to be handled with care. The other disadvantage of the Pearson correlation coefficient is that it is sensitive to outliers, and we have seen many of them in the scattergraphs in this book. This can work in two ways. A small group of outliers can exert a pull on the correlation coefficient that can result in a significant value for *r*, or the outliers may introduce 'noise' that can prevent us from seeing what should be a significant correlation. As noted above, a simple scatterplot between the two variables should allow any obvious outliers to be identified.

Exploring relationships between indices; regression

Regression goes one step further than correlation. Correlation can help tell us if two indices are linearly related to each other, albeit with no evidence for a direct cause–effect between the two, while regression tells us something about the nature of that relationship. Regression can tell us something about how one index changes as a result of another. Figure 9.2 is an example using the HDI data from Chapter 3 presented as Figure 3.3. Along the horizontal axis we have the HDI values for 1990 and the vertical axis comprises the HDI for 2017. Each dot in the graph is a single country. It was noted in Chapter 3 that there seemed to be a relationship between the two values of the HDI, such that countries having low HDIs in 1990 also have low HDIs in 2017, and the same is true for countries with medium and high HDI in 1990 and 2017. The point was made in Chapter 3 that this suggests a degree of durability with regard to the HDI rank; the relative

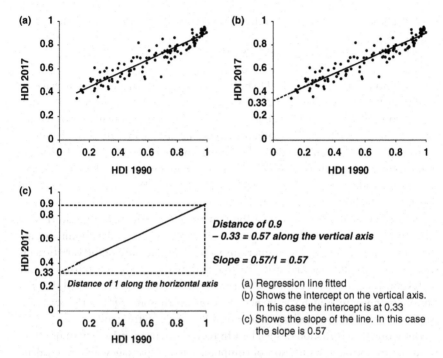

FIGURE 9.2 The fitting of a regression line to HDI data from 1990 and 2017

Source: Own creation using HDI data from the Human Development Reports 1990 and 2017.[1]

ranking of countries in terms of the HDI has stayed much the same over the 27 years between 1990 and 2017. We can calculate the correlation coefficient for these data to be 0.94, which gives us some statistical support for the assumption. But it was also noted in Chapter 3 that the data suggest an overall increase in values of the HDI for 2017 compared to 1990, suggesting that in general while their place along the HDI axes may not have changed that much, some countries had improved their level of human development over the 27 years. We can investigate this shift in HDI by fitting a regression line to the data as shown in Figure 9.2a. The line represents the best fit between all the data points. The approach taken here is to place a line through the data so as to minimise the vertical spread between all the data points and the line. Note the emphasis is upon vertical distances between the points and the line. The best fit line tells us two things. Firstly, we can see where the line, if extended (or extrapolated) towards the left-hand side, will cross the vertical axis (HDI 2017). In Figure 9.2b we can see that the extension of the best fit line crosses the vertical axis at 0.33. We can also estimate the slope of the best fit line, as shown in Figure 9.2c, and this tells us by how much HDI 2017 increases relative to HDI 1990. In fact, if we take the complete length of the horizontal axis as 1 then the corresponding distance travelled along the vertical axis is 0.57.

We can now write the relationship between HDI 1990 and HDI 2017 as:

HDI 2017 = 0.33 + 0.57 × HDI 1990

For a country with an HDI in 1990 of 0.2 we can expect that country to have an HDI in 2017 of 0.33 + 0.57 × 0.2 = 0.44, while for an HDI in 1990 of 0.8 we can expect an HDI in 2017 of 0.79. Thus the best fit line suggests that countries with lower HDIs in 1990 show a marked increase in 2017 while those with higher HDIs in 1990 have values in 2017 that remain more or less the same. In regression parlance we say that the HDI 2017 is dependent on the HDI 1990. Indeed, we call the index in the vertical axis (HDI 2017) the dependent variable while the one on the horizontal axis (HDI 1990) is the independent variable.

We can also calculate how well the best fit line fits the data. We do this by estimating how much of the variation we see in the scatter of data points is captured by our best fit line. The higher this value then the more of the variation (as seen in the two-dimensional space represented by the graph) is captured by our regression line. It turns out in this case with HDI 2017 and HDI 1990 that the capture is approximately 88%. Hence, we can explain some 88% of the variation we see in these data points by our best fit line; this is pretty good.

But while the estimation of such best fit regression lines is relatively straightforward and can be done with a host of computer software packages, and even calculators, these days, it does have its drawbacks. First, it is well to remember that the line we have fitted only applies to the swarm of data points we used to calculate it. Extrapolating either side of that swarm can be dangerous. In our HDI example the data points span most of the space we see in the graph, but at the low end (left-hand side) the last data point is 0.116, which is the HDI in 1990 for Niger. We can use our regression line to predict what the HDI in 2017 might be for a country having an HDI in 1990 that was lower than 0.116 but it would only be a prediction, and as we have seen over recent years, particularly following the financial crash in 2008, predictions of all kinds can be wrong as well as right.

Second, a regression line like this is based on an attempt to minimise the variation of points around it. Hence if we estimated the spread of the data points around the line in Figure 9.2a it would be the lowest we could get, at least with a straight line. Unfortunately, while this is mathematically easy to do, as seen above, it can cause problems if the distribution of data points in the swarm is uneven. In Figure 9.2a, for example, it would seem that at high values of the HDI (i.e. those more developed countries towards the top-right hand side of the graph) the points are closely packed and all of them are above the best fit line. Thus, the distribution of points is not regular across the data swarm, and we can see differences between those points at the high end of the scale compared to the rest. We may lose confidence that a one-line-fits-all approach such as we see in the graphs is merited.

Third, and related to the point above about differences in distribution of points along a best fit regression line, the use of such a best fit approach can also be influenced by groups of outliers; seemingly rogue data points that seem to sit away from the main swarm of points. The mathematics involved in fitting the best fit line does not discriminate; it treats every data point in the same way. While that is very egalitarian, it does mean that outliers can exert a significant influence on the best fit line, which can be quite large. But while they may be seen as disruptive and inconvenient, outliers have to be handled with great care. Eliminating them without good reason is dangerous as it leaves us wide open to claims of bias and fiddling the figures!

In the next section we will use correlation and regression to explore the relationships between the indices covered in this book. Indeed, in previous chapters we have already seen various scatterplots that seek to show the relationships between indices, but these were typically exploring changes in the values of indices over time. The next question is almost too obvious: What are the relationships between the indices? In cause–effect terms it is not difficult to come up with rationales by which every index in this book could be related, although as noted earlier these need to be handled with great care. It has to be stressed that there are many academic journal papers that do this, and with many other indicators and indices not covered in this book, in great detail, and these are far too numerous to cover here. Instead, I will opt to present just a few to give the reader a taste of the process and its pros and cons.

Relating indices: Does money rule?

Perhaps an obvious place to start is to see how the various indices in the book relate to income, expressed as GDP/capita. A series of scatterplots exploring this can be seen in Figure 9.3. As discussed in Chapter 2, there are various 'flavours' of GDP/capita that can be used, but here I have opted for current (2016) GDP per capita adjusted for purchasing power parity. In each case, I have included the correlation coefficient and shown the best fit linear regression line, although I have left out the regression equation.

The graphs in Figure 9.3 are based on a number of questions:

Does money buy you happiness? This is a question often asked, but it has a variety of answers, However the evidence presented in the scattergraph in Figure 9.3a suggests that maybe it does. The graph suggests that the Happiness Score increases with GDP/capita, and the correlation coefficient is 0.83, which is statistically significant at the 5% level.

Is corruption related to wealth? This question is a more specific variant of the question asked by Francis Galton (Galton, 1890), and the graph in Figure 9.3b suggests that this may also be the case, with a correlation coefficient of 0.73, which is significant at the 5% level. The poorer the country (lower GDP/capita) then the greater the level of perceived corruption using CPI 2015. This does have a logic. In poorer countries, perhaps with fewer checks and controls, one can readily imagine that officials with power will seek to make use of that to generate income.

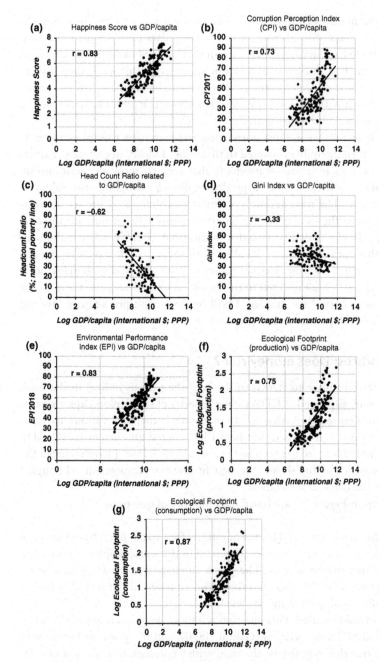

FIGURE 9.3 Series of scatterplots showing how various indices in the book relate to income (GDP/capita, adjusted for purchasing power)

Note: Each scatterplot has the correlation coefficient (*r*) and a 'best fit' linear regression line has been plotted through the field of dots. The regression equation has not been included.

Is the extent of poverty greater in poorer countries? This might sound like an obvious yes – after all, by definition poverty should be greater in poorer countries – but much depends upon how poverty is assessed. If we use the headcount ratio as the measure of poverty, then there is no theoretical given that a poorer country, as measured by GDP/capita, will have a greater proportion of its population below the poverty line. A country could, in theory, have a low mean GDP/capita but with almost all its population above the poverty line; in effect the mean could be pulled up by a minority of wealthy people with high levels of income and expenditure. The same could happen in reverse. A country might have a large disparity in terms of wealth, which generates a very high mean GDP/capita, but also have a high headcount ratio. As in the previous example, the high levels of expenditure from the wealthy are compensating for the lower levels of expenditure from the poorer segment of society. A mean is, after all, a mean, and it can hide a lot of variation. That is precisely why scientists and others always provide measures of variation alongside the mean. As shown in Figure 9.3c, with the headcount ratios based on the national poverty line, it does indeed seem as if poverty is linked to GDP/capita with the result that poverty is higher in poorer countries. The correlation coefficient is 0.62, which is significant at the 5% level.

Is inequality greater in poorer countries? A plot of the Gini Index against GDP/capita is shown in Figure 9.3d and, by and large, it is; countries with a higher level of income tend to be more equal in terms of how that income is distributed, with a correlation coefficient of −0.33, which, although seemingly low, is statistically significant at the 5% level. Higher values of the Gini coefficient relate to greater inequality in income. But note how the plots are more clustered than they are in the previous graphs, and the correlation coefficient is also lower. The Gini Index in this dataset only varies between 0.2 and 0.65. The evidence does not feel as compelling as it does for the other three graphs.

Is environmental performance related to income? There are various ways that this could play out. For example, the wealthier countries may have become rich at the expense of their natural environment so maybe the EPI would decline with wealth. The hypothesis here is that wealth was gained through activities such as mining or perhaps through factories that emit pollutants. Alternatively, and by way of a complete contrast, it could be argued that wealthier countries are able to commit resources to limit environmental damage through technology or perhaps through better enforcement of regulations (i.e. more resources committed to inspections). It could also be the case that wealthier populations are freed up from a need for basic survival and are able to consider the quality of the environment as being important and worth enhancing. The converse of this is that in poorer countries the emphasis of people is upon survival, even if this means cutting down trees or doing other things to degrade the environment. Figure 9.3e does indeed suggest that this last is

the case. It does indeed seem to be the case that the EPI for 2018 increases
with GDP/capita, and the correlation coefficient is a very respectable 0.83.

Is our footprint on the planet related to income? While the EPI attempts to assess the
environmental performance of nation states, what about the impact that
states are having on the planet? In Chapter 4 we looked at the ecological
footprint suite of indices, and the point was made that, while these have
been criticised over the years from a variety of angles, the notion of us having a 'footprint' on the planet is a powerful one that continues to resonate.
Perhaps of all the indices discussed in the book it is the one that seems
more tangible and relatable. But how does the EF relate to GDP/capita?
Figures 9.3f and Figure 9.3g are scatterplots of EF production and EF consumption respectively against GDP/capita and the broad message seems to
be that both forms of EF increase as GDP/capita increases. In other words,
increasing wealth means increasing footprint on the planet, and this is probably not all that surprising as a conclusion.

So, the evidence is Figure 9.3 appears to be compelling. Wealth really does play a
major role in explaining may of the index trends we see in this book. While we
can imagine scenarios where happiness, corruption, poverty, inequality, environmental performance and footprint on the planet were not related to GDP/
capita, and no doubt the scatter we can see in the graphs either side of the regression lines in Figure 9.3 is due to those scenarios on a country-by-country basis,
the overall pattern is what we would expect to see.

Before we move on to look at other comparisons between indices, I should
make a confession. The GDP/capita figures in the graphs in Figure 9.3 have all
been 'massaged' to help bring out the more linear relationships we can see with
GDP/capita. In each case, the massaging involved taking the natural logarithm
(LN) of the raw GDP/capita data; that is why the horizontal axes have been
labelled as log GDP/capita'. Natural logarithms are those using base e (where e
is approximately equal to 2.718281828459) rather than the more familiar base
10. As noted in Chapter 4, where I did something similar with the ecological
footprint and biocapacity scattergraphs in that chapter (and also in Chapter 3
with the construction of the Human Development Index), taking the logarithm
of data has the effect of compressing higher values and was done here because
the variation in GDP/capita across countries is very large indeed. For example,
Table 9.1 shows two values of GDP/capita, for Afghanistan and the US, and
their respective logarithms. The US has a GDP/capita that is more than 30 times
larger than that of Afghanistan; a very big difference indeed. However, taking
the logarithm of these two figures yields 7.5 and 11 for Afghanistan and the US
respectively, and the figure for the US is then only 1.5 times higher than that
of Afghanistan. The massaging (technically referred to as 'transformation') has
been carried out in a consistent way for all the countries. Indeed, for the ecological footprint graphs I have had to go further and take the natural logarithm of
the ecological footprint per country as well.

TABLE 9.1 The effect of using logarithms

Countries	GDP per capita, PPP (current international dollar) 2016	Logarithm (base e) of GDP per capita, PPP (current international dollar) 2016
Afghanistan	1,876	7.5
United States	57,466	11

Note: The range between high and low values is greatly reduced when logarithms are taken. GDP/capita PPP data have been taken from the Human Development Report of 2017.

Source: The GDP/capita figures were obtained from the World Development Indicators website of the World Bank (http://datatopics.worldbank.org/world-development-indicators/).

One other aspect of using logarithms is that it can simplify regression equations. This is because multiplications and divisions become additions and subtractions respectively. We can illustrate this advantage using logarithms to the base 10 (the same principles apply to natural logarithms). For example, in Chapter 3 we saw that the logarithm (base 10) of 10 is '1' and of 100 is '2'. Multiplying 10 by 10 gives 100, and this is equivalent to adding together the logarithms of 10 such that $1 + 1 = 2$. Similarly, 100 divided by 10 is 10, or using logarithms, $2 - 1 = 1$. Thus, if we use logarithms then multiplication and division become much easier as the equations simplify to additions and subtractions. This was the principle of the slide rule, an early form of mechanical calculator that long predates the electronic devices we have today.

The use of logarithms is a widely applied and useful way of allowing a simple linear regression to be fitted and doing this provides the kind of neat-looking trends we see in Figure 9.3 together with the possible explanations I have provided for each of them. They allowed me to produce graphs similar to those in 9.1a and b and give the correlation coefficients so the reader could get a sense of statistical evidence for each of the relationships rather than just relying on the patterns of dots. However, using transformations can also be deceptive. I will illustrate the point with the EPI and ecological footprint data used to comprise Figures 9.3e, f and g. In Figure 9.4 the graphs have been produced with the raw data, with no logarithms being used for the GDP/capita or EF and EPI data, and the patterns do look different from the equivalent graphs in Figure 9.3. For the EF graphs, what is perhaps most interesting is the spread between data points at the left- and right-hand sides of the graphs. There is a group of countries bunched together at the low end of the EF and GDP/capita axes but the points tend to spread out as GDP/capita increases. At any particular GDP/capita, for example US $50,000/capita, there is a marked difference in EF across countries. In the case of EF based on consumption, the spread is from just over 5 up to 9 global hectares per person, and for EF production it is even greater – from 4 to nearly 14 global hectares/person. It seems that some countries are able to have the same wealth as others but achieve this with a much lower footprint (impact) on the planet. Herein rests an important principle; the results tell us that it is possible for

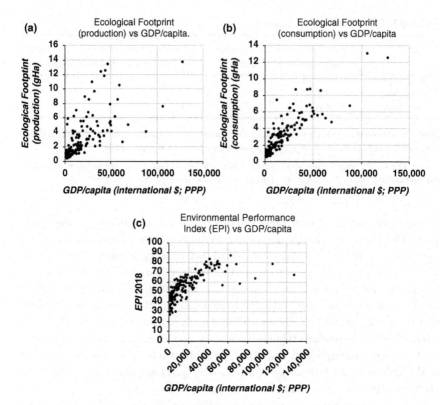

FIGURE 9.4 Scatterplots showing the EF and EPI plotted against GDP/capita (adjusted for PPP)

Note: In this case none of the data have been transformed using logarithms.

a country to maintain its level of wealth but reduce its footprint on the planet by what is potentially a significant margin. Admittedly the achievement of such a shift may take a lot of effort both in terms of education and in encouraging different behaviours of consumption and production, but it can be done.

For the EPI, the graph with the raw GDP/capita values brings out an interesting and important feature of the relationship. We lose the simple linear relationship seen with the logarithms and replace it with a curve that suggests we have a diminishing return. As GDP/capita increases, we do indeed have an increase in the EPI but only up to a point. The relationship begins to flatten out, suggesting that increasing wealth does not necessarily bring better environmental performance. But why should that be the case? Why does environmental performance level off like this? First, we have to keep reminding ourselves that the EPI is a human construct that tries to capture environmental performance. Hence it is quite possible that what we are seeing is an artefact caused by the nature of the EPI. Maybe if we measure environmental performance in different ways then the relationship we observe with GDP/capita will also change. There are also natural ceilings/floors in data. For example, it is obviously not possible to get a figure of more than 100% of a population having access to sanitation or exposure to air pollution or less than 0% of waste water being treated. Changing the definitions of these terms, all of which do appear in the EPI, will change the values for any particular country of course, but, however they are defined, the ceilings and floors for the percentages remain at 100% and 0% respectively. Thus, we can expect a degree of bunching of country values around the ceilings and floors provided by components of the EPI. But beyond these technical arguments of index construction can we not also come up with some other possible explanations for such a diminishing return? There are indeed many possible reasons. For example, governments have many priorities for the allocation of limited funds and, as their wealth increases, they may well opt to spend money on other things besides the environment. Indeed, there may be a tendency to address environmental concerns up to a point that shows they are doing something, but not going any further than that. There is also the possibility that we are limited by what technology can achieve at any one time. Cars have become much 'cleaner' over the years when it comes to emissions of gasses and particulates, and the new generation of electric cars has produced cars with almost no emissions, but new technology comes at a cost and can take time to be widely adopted even within wealthy countries.

Carried away

As the reader can probably sense from the previous section, it is all too easy to get carried away when working with indices. It is intriguing to ask whether indices are related and, if so, what may be the possible causal mechanisms at play. Added to that, the creation of scatterplots, correlations and regressions shown so far in this chapter are all very straightforward, with widely available software such as Microsoft Excel. It literally takes seconds to generate these plots once the data have been input. Even generating the best fit regression lines and statistics is but a right-click of a mouse button away. As already mentioned, the academic literature that attempts to relate indicators and indices to each other, over time and space, is vast and cannot possibly be covered here. However, in such an easy process there are dangers, of course, that the reader should be aware of; and here are just a few of them.

First, it can all become too mechanical, meaning that we can lose sight of what we're trying to achieve. The graphs should only be a means to try to understand what is happening in the data and should not be the be-all and end-all. It is also important to note that using correlation and regression does not tell us anything about the mechanisms involved; just how is the independent variable influencing the dependent one? Indeed, are the two variables the right way round? Is it right to think of EF as a product of wealth or should it be the other way round – that countries get wealthy by having a high impact on the planet? The analyses can be run either way and the software does not make that decision for us; the software is entirely neutral in that key decision.

Second, the use of such best fit regression comes with various technical assumptions. One of these is that the distribution of points around the best fit line should be more or less the same along the length of the line. If the extent of scatter around the line is larger at one end on the line than the other then this generates what statisticians called heteroscedasticity (some also call it heteroskedasticity); a word taken from the ancient Greek with 'hetero' meaning 'different' and 'skedasis' meaning 'dispersion'. A good example of this can be seen with the two ecological footprint graphs in Figure 9.4 and indeed discussed above. The dots are far more bunched towards the left-hand side of the graphs, while towards the right-hand side the variation between points for the same GDP/capita is much larger. Fitting a regression line to these graphs, while entirely possible and easy to do with software, would be misleading and I have not done it. Indeed, this is another reason why I took the logarithms of the two variables and fitted the lines to those in Figures 9.3f and g. Some more specialised computer software designed to run statistical tests will often flag up issues such as heteroscedasticity, among other potential problems, but it is up to us to notice such flags and make decisions over what to do.

Third, extrapolation of such best fit lines is tempting but risky. The regression line is based on the data we used to create it and there are no guarantees that the relationship would apply to data that are outside of that range. That is why the fitted lines you can see in Figure 9.3 are within the data points and do not extend outside of them. If we choose to extrapolate the line to make predictions of dependent variables, then we do so at some risk.

Finally, all the analyses in the previous section had GDP/capita (adjusted for PPP) as the dependent variable. The impression being given is perhaps that GDP/capita is the key indicator with all the others in the book revolving around it; just as planets of the solar system orbit the Sun. But given that all these indices are related to GDP/capita then they will also be related to each other. Rather than show all the graphs here, we can in fact summarise this using a matrix of correlation coefficients as shown in Table 9.1. The table lists all the indices in the columns and rows and the numbers are the correlation coefficients. Negative values for the correlation coefficients mean that as one of the indices increases the other declines, while positive values mean that as one increases then so does the other. Each index is perfectly correlated with itself, of course, and that is why you can see the diagonal lines of 1s. All the correlation coefficients are statistically significant (at the 5% significance level) except for three of them, shaded in the table, which are for comparisons involving the Happy Planet Index. Two of the non-significant comparisons are between the Happy Planet Index and the ecological footprint (consumption and production) while the third is between the Happy Plant Index and the Gini Index. This might be surprising to the reader as, after all, the HPI contains both the EF and the Gini coefficient, but these are just two elements of the four that go to make up the HPI. Clearly, the inclusion of the other two components, for happiness and life expectancy, is enough to break a linkage that we might assume exists.

Table 9.2 gives us the bigger picture, part of which we have already seen with the graphs included in this chapter. The message is perhaps an unsurprising one: Most of the indices are related, although, as we need to keep reminding ourselves, there is nothing in here that tells us why or how they may be related. Indeed, the coefficients do not tell us which of the indices is dependent on the other. For example, the table tells us that poverty is related to corruption, but does poverty cause corruption or does corruption cause poverty? It is probably a bit of both but the correlation coefficient tells us nothing more than that, as one index (corruption) increases in size (i.e. corruption gets less), then the other index (headcount ratio – national and international poverty lines) also declines (i.e. poverty declines). Similarly, as the Environmental Performance Index (EPI) increases then poverty worsens; but is poverty causing the decline in environ-mental performance or are people becoming poorer because the environment is degrading? Is it not a self-reinforcement where each component is leading to a decline in the other – poverty leads to environmental degradation, which in turn leads to more poverty? These are complex questions; far more complex than a simple table of correlation coefficients suggests. We can think of mechanisms for all these relationships, but this requires far more thought and probably more evidence than it took to generate the table.

Table 9.2 and indeed the scatterplots in this chapter are all illustrations of the large amount of effort that has gone into exploring how indices can help explain important relationships in local and global societies. Indeed, the everyday media are replete with such analyses and stories, and there are so many indicators and indices out there, covering just about every facet of life that you can think of, that means indicator and index geeks like myself almost have an embarrassment of riches. Indeed, that is arguably one of the problems faced by indicator/index geeks like me; it is so easy to get carried away with generating these tables and graphs that we can lose sight of the deeper messages. The world, and all its rich-ness and complexity, becomes reduced to a list of numbers and a few graphs.

But while it is easy to get carried away there are always dangers that lurk around the corner. There are always concerns about the quality of the data used in these analyses, of course, as well as the numerous assumptions behind each index, as we have seen in these pages. It is so easy to paint the world in a way that consciously or unconsciously reflects one's own biases, but that is given a façade of legitimacy by the use of numbers and statistics. All this arguably plays to an almost basic instinct we seem to have in wanting to simplify the world around us so we can grapple with it. Indeed, it is tempting is it not to create an 'index of indices' – one index to rule them all (using a Mordor-esqe metaphor). Can we not roll up environment performance, ecological footprint, human devel-opment, as just a few examples, within a bigger 'meta-index' that captures all aspects of human life within each country? Why should we stop at listing each of the indices, especially as each of them in turn also has many components? If the HDI uses education, healthcare and income to capture 'human development', then why cannot we go one step further than that and pull in environmental

TABLE 9.2 Table of correlation coefficients (r) for all the indices discussed in this book

INDICES and year	LN GDP/ capita 2016	HDI 2017	Happiness score 2017	HPI 2016	CPI 2017	EPI 2018	LN of EF Production	Consumption	Headcount Ratio International	National	GINI index
LN GDP/capita 2016	1	0.95	0.83	0.29	0.73	0.83	0.75	0.87	-0.78	-0.62	-0.33
HDI 2017		1	0.84	0.38	0.74	0.86	0.72	0.83	-0.79	-0.70	-0.37
Happiness score 2017			1	0.50	0.70	0.79	0.71	0.75	-0.66	-0.55	-0.29
HPI 2016				1	0.19	0.30	0.02	0.00	-0.45	-0.39	-0.07
CPI 2017					1	0.72	0.63	0.69	-0.39	-0.46	-0.34
EPI 2018						1	0.64	0.75	-0.64	-0.52	-0.35
LN EF (production)							1	0.90	-0.53	-0.52	-0.31
LN EF (consumption)								1	-0.63	-0.56	-0.35
Headcount ratio (international)									1	0.72	0.26
Headcount ratio (national)										1	0.40
GINI Index											1

Note: For three of the indices, GDP/capita, EF (production) and EF (consumption), the correlations are based on the natural logarithm (LN) of the raw data.

The three lightly shaded cells are those of correlation coefficients not statistically significant at the 5% (equivalent to 1 in 20) cut–off point that is often used by convention with such statistical tests. Note how all three of the non–significant correlations include the HPI.

performance and impact on the planet to provide a single index for each country? It is a compelling path, but where should such integration stop?

Integration of indicators can be achieved in various ways, one of which has been shown with almost all the indices described in this book – by taking the average! Admittedly the various indices have been based on assumptions regarding weighting of components and transformation, but adding them up and dividing by the number of components is just about the most straightforward way of achieving aggregation. The Human Development Index, for example, has, since birth, had three components that are added and divided by three. It does not get more straightforward than that. The Environmental Performance Index, and its predecessor the Environmental Sustainability Index, both have a far more complex aggregation process, with all kinds of weightings for numerous components, but in essence, that is all there is to it. There are more complex approaches to integration, of course, and an example is provided by the Happy Planet Index, which is a ratio having three components on top (comprising the numerator) and one on the bottom (denominator). The three numerator components are added together to, in effect, provide the 'benefit', while the denominator is the ecological footprint, representing the cost. Thus the HPI is, in essence, a benefit:cost ratio, expressed in non-monetary terms, showing how much benefit we receive for each unit of cost. It is the only such example provided in this book and it may be argued that one of the reasons for its resonance with policymakers is that sense of cost associated with benefit. One can do the same, of course, for many of the other indices in this book – and researchers have done just that. For example, we could divide the HDI by the ecological footprint (consumption) to get a sense of gain in HDI per unit of EF. This has been something of a popular approach among indicator experts, and, as with the HPI, the idea is to think of gains in development relative to an impact on the planet. Figure 9.5a shows how the two variables are related and, as we would probably expect from some of the earlier discussions, we do see a levelling of the HDI as the EF increases; there are many countries having high values for the HDI but wide variation in their impact on the planet. How this benefit:cost ratio varies across the globe is shown in Figure 9.6 and there is a clear north–south divide. The best returns on HDI (the benefit) per unit of the ecological footprint (the cost) are to be found in Africa and parts of Southern Asia, while the worst returns are to be found in the developed world. But, of course, having a high benefit per cost does not necessarily equate to a country having a desired level of development. Many of the countries with the best HDI:EF ratios are also those with the greatest degree of poverty, as illustrated in Chapter 6.

We can take the same approach for other indices, of course, and Figure 9.5b relates the Environmental Performance Index to the EF. We can think of this, ironically, as the environmental cost of environmental performance! As we can see in the graph, a very similar pattern to that of the HDI:EF emerges with better EPI coming at a greater cost in terms of EF. The reason, of course, is that the EPI has components that are related to financial cost, and thus what we are seeing in

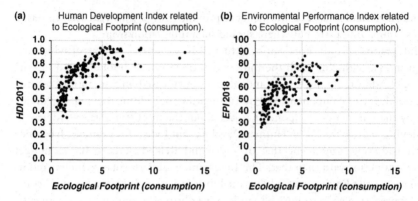

(a) Human Development Index related to Ecological Footprint (consumption).

(b) Environmental Performance Index related to Ecological Footprint (consumption).

FIGURE 9.5 Human Development Index (for 2017) and Environmental Performance Index (for 2018) related to the Ecological Footprint (consumption)

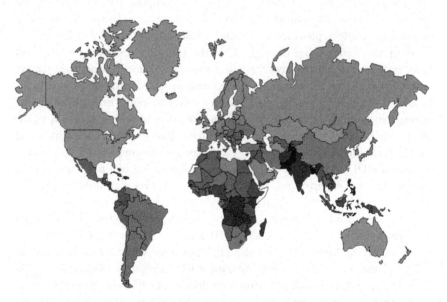

FIGURE 9.6 Global map showing the ratio between the HDI as a benefit and EF as the cost. Darker shading equatez to higher values for the benefit:cost (HDI:EF) ratio and imply greater returns on HDI per unit EF

Figure 9.5b is probably an undertone of wealth. Nonetheless it illustrates the care we need to take when making such linkages. Are we really looking at benefit expressed by one indicator and cost by another or are the two really reflecting a third dimension such as economic performance?

While I have used some simple statistical methods in this chapter, essentially no more than correlation and regression, there are more powerful methods in the armoury of those who work with indicators, including, for example, more sophisticated regression techniques and factor analysis. Factor analysis begins with the table of correlation coefficients shown in Table 9.1, but goes much further in teasing apart the relationships between the indices. There is no space to go into all of these here and the interested reader is referred to introductory texts such as Kline (1993). I will only provide one intriguing insight here, and this is derived from cluster analysis, a relative of factor analysis designed to help identify categorisations or clusters. The starting point for this example of cluster analysis is to use the correlation coefficients shown in Table 9.1, and the output is the dendrogram in Figure 9.7. The indices are listed along the horizontal axis, and the vertical axis represents similarity. It is almost like a family tree that starts at the top of the diagram and then, based on closeness of apparent relationship, identifies clusters of more closely related family members towards the bottom of the diagram. The dendrogram suggests that there is a cluster of indices comprising GDP/capita, the HDI, the EPI, the Happiness Score, the EF and the CPI that all seem to have a strong relationship with each other. We can call these closely related indices Group A, and it would be interesting to explore why they are so closely related. Maybe, as intimated in this chapter, it is wealth that truly binds them together, despite the best endeavours and wishes of their creators. There is a second grouping towards the right-hand side comprising the headcount ratio and Gini Index, which we can call 'Group B'. It is easy to see why the two headcount ratios should be related. In between these two groups sits the Happy Planet Index. While the HPI does have more in common with Group A than Group B it has split from Group A and seems to be more on its own. Maybe this is because, unlike all the others in the diagram, it is an index that is built on benefit:cost.

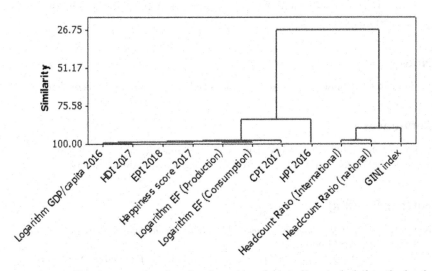

FIGURE 9.7   Dendrogram showing the clustering of the indices included in the book. The indices are shown across the horizontal axis while the vertical shows the degree of similarity between the clusters

Needless to say, the academic literature on indices and indicators is replete with studies of the type I have outlined in this chapter, comparing them across all sorts of units such as national states and companies as well as looking for trends over time. Care needs to be taken when undertaking such analysis for all the reasons I have outlined, but we can learn a lot about ourselves including the thinking that went behind the construction of the indices. The sky is the limit.

Conclusion

Having lots of indices at our fingertips leads to the strong temptation for us to see how they may be related to each other. There are so many questions we can ask, and this chapter has sought to provide the reader with a taste of some of these as well as looking at how any such relationships can be explained. We can learn a lot from how indices relate to each other and indeed the variation that we can see between countries having similar values of one index but widely different values of another. Why should these countries be so different?

The issues raised in this book about consistency of index methodology, quality and quantity of data used, assumptions behind indices, etc. all need to be born in mind with such work, but the rewards in terms of helping us understand the human condition are great. So much has been done but there is so much still to do.

Finally, it will not have escaped the reader's attention that this chapter in particular, as well as the book so far, has concentrated on indices that are numerical. Almost by definition we have assumed that all indices must be numerical; they must be numbers. Numbers have had a powerful hold within the human psyche and it is often said that you cannot manage what you cannot measure. But there are other indicators out there that are not numerical and these will be the focus of the next chapter.

Note

1 Human Development Reports are available at: http://www.hdr.undp.org/en.

References

Galton, F (1890). Kinship and correlation. *North American Review* 150, 419–431.
Kline, P (1993). *An Easy Guide to Factor Analysis*. Routledge, Abingdon, UK.

Further Reading

Rowntree, D (2000). *Statistics without Tears: An Introduction for Non-Mathematicians*. Penguin Books, London.
Urdan, T C (2017). *Statistics in Plain English*. 4th edition. Routledge, New York and London.

10

WHERE ARE WE GOING?

Introduction

In this final chapter of the book, I want to discuss the future of indicators and indices. Part of this future story has already been written via global-scale programmes such as the Sustainable Development Goals (SDGs) established by the United Nations.[1] The SDGs have an associated set of indicators, more than 300 in number, and the many countries that have signed up to the SDGs have to report progress on these until 2030. Thus, at a global scale, the SDG indicators will be with us for at least another decade, whether we like it or not, and the first part of this chapter will present that agenda and the development of one of the newest indices, but one that will no doubt grow in importance, the Sustainable Development Goal Index (SDGI).

But what are the other frontiers in the indicator world? There are many answers to this question, and almost everyone who works in the field will have a different view. Indeed, having read the other chapters of this book, the reader may well have developed their own answers to this question. For example, one field where I feel much more work could be done is in the *use* of indicators and indices. Who uses them, for what purpose and how? Throughout the book we have seen indices created by people to promote concerns to a wider, often nonspecialist, audience and provide a way in which progress (or lack of it) can be assessed. But to what extent have they succeeded in achieving change? Following on from this, how do users think the indices could be changed or perhaps presented to make them more usable and hence impactful? These are some obvious yet 'big' questions to ask, and, while research that helps to provide some answers does exist, much more could be done. The reader interested in some possible answers to these questions is referred to the edited book by Bell and Morse (2018). In this chapter, rather than focus on this 'use' frontier for indices,

I will instead look at the contribution from a technology that I feel is growing in potential for indicators – Earth observation via satellites.

Scoring goals

Goals seem to have been something of a fashion in the indicator world since 2000.

In 2000, the United Nations established what it called the 'Millennium Summit', which sought to address many of the important issues affecting people's lives in the developing world. The summit set out the Millennium Development Goals (MDGs), which countries were asked to sign up to and achieve by 2015. The MDGs covered a range of social, economic and environmental challenges faced primarily in the developing world and a suite of targets, as well as corresponding indicators to measure progress towards the targets, was established for the various goals. These were intended to be routinely measured by national governments and reported back to the UN.

There were eight MDGs in total and these can be divided into three groups: 'social', 'environmental' and 'global partnership for development'. Table 10.1a lists the more social MDGs and associated indicators while Table 10.1b lists what are called the environmental sustainability and global partnership for development goals. Table 10.1 comprises a rather long list of indicators and I do not want to go through each of them here. The reason for including the list in its entirety is to allow the reader to make the links back to many of the same sorts of issues covered in other chapters of this book. In Table 10.1a, the social goals, there are the usual suspects one would expect to see in a list of this sort, including GDP-based indicators, education and healthcare, as well as poverty, gender equality and empowerment. Much of the list is not surprising and indeed we have already touched upon the issues in the chapters on GDP, HDI and poverty (Chapters 2, 3 and 6). In Table 10.1 there are elements in the environmental list that resonate with issues discussed in the chapter on the ESI and EPI (Chapter 5), such as the proportion of land area covered by forest, species being faced with extinction, availability of improved drinking water and sanitation. The global partnership for development theme also has resonance with ideas in the ESI that were focussed on institutions, partnerships, etc. required for the implementation of environmental sustainability. The list of MDG indicators reminds us that the challenges we face in development have been remarkably persistent over many decades; something that we have caught a glimpse of with the consistency in country ranking we see with indices such as the HDI. Different indicators have been created and evolved to help us assess progress towards addressing the issues, but the issues have a disturbing degree of persistence in the face of all our attempts to address them. Unfortunately, the development of new indicators and indices may help with defining these issues and knowing what we need to focus upon, but it does not make them go away; if only it were that easy.

TABLE 10.1 The social development components of the MDGs

Millennium Development Goal	*MDG indicators*

(a) The social components of the World Development Goals and their indicators

Eradicate extreme poverty and hunger	Proportion of population below $1 purchasing power parity (PPP) a day
	Poverty gap ratio [incidence × depth of poverty]
	Share of poorest quintile in national consumption
	Prevalence of underweight children under 5 years of age
	Proportion of population below minimum level of dietary energy consumption
	Growth rate of GDP per person employed
	Employment to population ratio
	Proportion of employed people living below $1 (PPP) a day
	Proportion of own account and contributing family workers in total employment
Achieve universal primary education	Net enrolment ratio in primary education
	Proportion of pupils starting grade 1 who reach last grade of primary education
	Literacy rate of 15- to 24-year-olds
Promote gender equality and empower women	Ratios of girls to boys in primary, secondary and tertiary education
	Share of women in wage employment in the non-agricultural sector
	Proportion of seats held by women in national parliament
Reduce child mortality	Infant mortality rate
	Under-5 mortality rate
	Proportion of 1-year-old children immunized against measles
Improve maternal health	Maternal mortality ratio
	Proportion of births attended by skilled health personnel
	Adolescent birth rate
	Antenatal care coverage (at least 1 visit and at least 4 visits)
	Unmet need for family planning
	Contraceptive prevalence rate
Combat HIV/AIDS, malaria and other diseases	HIV prevalence among population ages 15–24 years
	Condom use at last high-risk sex
	Proportion of population ages 15–24 years with comprehensive, correct knowledge of HIV/AIDS
	Ratio of school attendance of orphans to school attendance of non-orphans ages 10–14 years
	Proportion of population with advanced HIV infection with access to antiretroviral drugs
	Incidence and death rates associated with malaria
	Proportion of children under age 5 sleeping under insecticide-treated bed nets
	Proportion of children under age 5 with fever who are treated with appropriate antimalarial drugs
	Incidence, prevalence, and death rates associated with tuberculosis
	Proportion of tuberculosis cases detected and cured under directly observed treatment short course

(Continued)

TABLE 10.1 Continued

Millennium Development Goal	*MDG indicators*

(b) The non-social development components of the World Development Goals and their indicators

Ensure environmental sustainability	Proportion of land area covered by forest
	Carbon dioxide emissions, total, per capita and per $1 GDP (PPP)
	Consumption of ozone-depleting substances
	Proportion of fish stocks within safe biological limits
	Proportion of total water resources used
	Proportion of terrestrial and marine areas protected
	Proportion of species threatened with extinction
	Proportion of population using an improved drinking water source
	Proportion of population using an improved sanitation facility
	Proportion of urban population living in slums
Develop a global partnership for development	Net Official Development Assistance (ODA), total and to the least developed countries, as percentage of OECD/ Development Assistance Committee (DAC) donors' gross national income
	Proportion of total bilateral, sector-allocable ODA of OECD/ DAC donors to basic social services (basic education, primary healthcare, nutrition, safe water and sanitation)
	Proportion of bilateral official development assistance of OECD/DAC donors that is untied
	ODA received in landlocked developing countries as a proportion of their gross national incomes
	ODA received in small island developing states as a proportion of their gross national incomes
	Proportion of total developed country imports (by value and excluding arms) from developing countries and least developed countries, admitted free of duty
	Average tariffs imposed by developed countries on agricultural products and textiles and clothing from developing countries
	Agricultural support estimate for OECD countries as a percentage of their GDP
	Proportion of ODA provided to help build trade capacity
	Total number of countries that have reached their heavily indebted poor country (HIPC) decision points and number that have reached their HIPC completion points (cumulative)
	Debt relief committed under HIPC Initiative and Multilateral Debt Relief Initiative
	Debt service as a percentage of exports of goods and services
	Proportion of population with access to affordable essential drugs on a sustainable basis
	Telephone lines per 100 population
	Cellular subscribers per 100 population

Source: Own creation based on information available at http://www.un.org/millenniumgoals/.

Achievement of the MDGs by 2015 was patchy across countries (Andrews et al., 2015; Cimadamore et al., 2016); some (notably Brazil, India and China) did well while others not so. In 2016, the MDGs were replaced by 17 Sustainable Development Goals (SDGs), which followed the same pattern as the MDGs by having an associated framework of targets and indicators (Table 10.2). However, unlike the MDGs, the SDGs are intended to apply to all countries, not just those considered to be less developed, and thus require a global level of action with partnership between governments and agencies rather than the 'aid giver–recipient' dynamic set out in the global partnership for development category of the MDGs. The list of targets and indicators for the SDGs is a long one, and at the time of writing there are 169 targets and 230 agreed indicators for those targets. Countries are meant to use these indicators to help measure their progress towards the SDG targets, with the year 2030 given as the end date. I do not wish to list all the SDG indicators and targets here; the list for the much more restricted MDGs has already required two lengthy tables in this book.

Indeed, the long list of SDG targets and indicators can be both a positive and a negative. On the plus side, and as we have already discussed throughout the book, sustainable development is complex and multifaceted, and thus one must expect to have many goals, targets and indicators. The approach of trying to reduce this complexity to a single index can be oversimplistic, and the creators of the SDG framework have avoided that criticism. On the other hand, some have noted that the SDG framework is bewildering, and an article in *The Economist* refers to the framework as "sprawling and misconceived", "a mess" and potentially a "betrayal of the world's poorest people". *The Economist* (2015) even drew a parallel with the Old Testament:

> Moses brought ten commandments down from Mount Sinai. If only the UN's proposed list of Sustainable Development Goals (SDGs) were as concise.

Strong words indeed but the article does make a valid point. Being concise runs the risk of presenting an oversimplistic picture, but it has its advantages, and, as we have seen in the other chapters, using single indices rather than lengthy indicator frameworks can help with the presentation of complexity to those who are not necessarily experts in the subjects that are being condensed.

Another issue that can emerge with multifaceted and extensive indicator frameworks is the danger of contradiction; some indicators in the framework may work against others. This is a common problem, and indeed we have encountered it already even with the few indices covered in this book. A government, for example, may opt to prioritise growth in GDP even if that results in a higher ecological footprint or worsening environmental sustainability and performance. But if a single indicator framework has such contradictions embedded within it, and governments and others are being asked to achieve such targets, then the dangers can be exacerbated. In effect, progress in some targets may be compromised by progress in others.

TABLE 10.2 The SDGs: Number of targets per goal and number of indicators included in the SDGI 2018

SDG	Short title	Description	Number of targets	Number of indicators in SDGI	
				Global	OECD
SDG 1	No poverty	End poverty in all its forms everywhere	7	2	3
SDG 2	Zero hunger	End hunger achieve food security and improved nutrition and promote sustainable agriculture	8	6	6
SDG 3	Good health and well-being	Ensure healthy lives and promote well-being for all at all ages	13	14	17
SDG 4	Quality education	Ensure inclusive and quality education for all and promote lifelong learning	10	3	8
SDG 5	Gender equality	Achieve gender equality and empower all women and girls	9	4	5
SDG 6	Clean water and sanitation	Ensure access to water and sanitation for all	8	4	4
SDG 7	Affordable and clean energy	Ensure access to affordable, reliable, sustainable and modern energy for all	5	3	4
SDG 8	Decent work and economic growth	Promote inclusive and sustainable economic growth, employment and decent work for all	12	4	5
SDG 9	Industry, innovation and infrastructure	Build resilient infrastructure, promote sustainable industrialization and foster innovation	8	7	11
SDG 10	Reduced inequalities	Reduce inequality within and among countries	10	1	3
SDG 11	Sustainable cities and communities	Make cities inclusive, safe, resilient and sustainable	10	3	4
SDG 12	Responsible consumption and production	Ensure sustainable consumption and production patterns	11	7	7
SDG 13	Climate action	Take urgent action to combat climate change and its impacts	5	4	5
SDG 14	Life below water	Conserve and sustainably use the oceans, seas and marine resources	10	6	6

SDG 15	Life on land	Sustainably manage forests, combat desertification, halt and reverse land degradation, halt biodiversity loss	12	5	5
SDG 16	Peace, justice and strong institutions	Promote just, peaceful and inclusive societies	12	9	9
SDG 17	Partnerships for the goals	Revitalize the global partnership for sustainable development	19	4	5
			169	86	107

Source: Adapted from Lafortune et al. (2018).

One of the challenges faced by those having to work with the SDG suite of indicators, and indeed those of the earlier MDGs, is the need for both quantity and quality of data to populate them. This is a challenge as resources need to be committed to data collection and verification, and this can be expensive and time-consuming. Technology may help here and, in a later section in this chapter, we will explore one such aid, namely Earth observation via satellites and other technologies.

One index to rule them all, one index to bind them

In Chapter 9 it was noted that there is often a temptation to pool all the indices into one; the indicator world equivalent of the one ring to bind them all that we see in the *Lord of the Rings* books and movies. The logic is straightforward; if all the indices are correlated and if they span many of the issues important in sustainable development, then why not put them together and have a mega-index that tracks the progress of each country? In fairness, the creators of the SDG framework of indicators are aware of the criticisms that can be applied to such complex lists, and they have developed a single index – called, rather unimaginatively but accurately, the Sustainable Development Goal Index (SDGI). The idea, as with all the indices presented in this book, is to have a single headline number for each country as a means of showing progress towards attainment of the SDGs by 2030. It is perhaps the best example we have today of this desire to have a single index to rule them all.

The construction of the SDGI is straightforward and, in tune with all the indices covered in this book, the creators of the index have made a number of key assumptions and decisions. When faced with decisions, for the most part they have opted for simplicity rather than complexity. The starting point is a somewhat reduced list of indicators within each of the 17 SDGs. For the 2018 version, the number of indicators included in the SDGI is shown in Table 10.2. For all countries a total of 88 indicators were employed (37% of the total of 230 agreed indicators), but some additional ones (21) were added for countries that are members of the Organisation for Economic Co-operation and Development (OECD). The OCED comprises 36 of the more developed countries of the world, and the decision to add more indicators for OECD countries appears to have been based on the availability of more data of good quality for those countries. Hence, while the SDGI has much in common with the indices covered in this book, it differs from most of them in being a construct of very many indicators. In that sense it resonates a little with the ESI and EPI covered in Chapter 5, both of which were also indices comprised of many indicators. For example, the ESI included a total of 65 variables in its 2000 version and 76 in its 2005 version, but the 88 global SDGI indicators and 107 for OECD countries exceeds even this.

The creators of the SDGI employed five criteria for making the selection of indicators from the long list (230) included in the SDGs, and these are set out as follows (using the words of the creators of the SDGI with emphases added by me):

1. *Global relevance and applicability to a broad range of country settings*: The indicators are relevant for monitoring achievement of the SDGs and applicable to the entire continent. They are internationally comparable and allow for direct comparison of performance across countries. In particular, they allow for the definition of quantitative performance thresholds that signify SDG achievement.
2. *Statistical adequacy*: The indicators selected represent valid and reliable measures.
3. *Timeliness*: The indicators selected are up to date and published on a reasonably prompt schedule.
4. *Data quality*: Data series represent the best available measure for a specific issue, and derive from official national or international sources (e.g. national statistical offices or international organizations) or other reputable sources, such as peer-reviewed publications. No imputations of self-reported national estimates are included.
5. *Coverage*: Data have to be available for at least 80% of the 149 UN Member States with a national population greater than 1 million.

(Lafortune et al., 2018, pp. 7 and 8)

The indicators are adjusted to reflect 'distance to target'; how far each indicator value is away from where is needs to be, with the targets for each indicator set by the UN. The distance to target approach for standardising variables to a shared scale is common and we have seen examples of it with other indices in this book, most notably the Human Development Index and the Environmental Performance Index. Please note that the target for each of the 88 (or 109 for OECD countries) indicators should not be confused with the 169 targets provided in Table 10.1 for each SDG; as we saw in Chapter 5, terminology with indicators and indices can be confusing. In the case of the SDGI the scale chosen is 0 (no progress at all towards target) to 100 (achievement of target). Once this has been done for all the indicators within an SDG the values are averaged (arithmetic mean) to create a value for each SDG. Once averaged to produce a single value for each SDG then it is suggested by the creators of the SDGI that this be interpreted as percentage attainment of goal. The creators of the SDGI purposely opted for the arithmetic mean rather than, for example, the geometric mean as used in more recent versions of the HDI, because they wished to keep the calculation, and hence interpretation, as simple and intuitive as possible.

If there are missing values for an SDG then the creators of the SDGI have simply used the average value for that SDG from the country's peer group of nations. The issue of missing data often emerges in index creation, as discussed in the other chapters of the book, but adding missing values in this way is certainly straightforward and is probably the simplest way of addressing the problem. An example is shown in Table 10.3 for Africa. The two SDGs that appear to be most affected by missing data are SDG 10 ('Reduced inequality') and SDG 14 ('Life below water'), and in Table 10.3 there are many missing values for those two SDGs across the African nations. Indeed, in some cases such as for Ethiopia, and Zambia, there are missing data for both SDGs. The average value for SDG 10 across these countries is 45.2 while for SDG 14 it is 48.2. These averages are then plugged into all the respective gaps in Table 10.3 so the SDGI can be calculated. However, while filling in gaps like this is straightforward the danger is that it is an oversimplification. For example, in SDG 10 the values range from 0 (Botswana and Namibia) to 86.3 (Mauritania), and the values are shown plotted in order in Figure 10.1, with the highest value to the left and lowest to the right. Does it seem reasonable to use the dotted line (average) of all countries with values to plug the gaps we see in the table for SDG 10? Why would Angola not be closer to the values for Namibia or Botswana, its near neighbours, both of which had values of 0 for SDG 10? On the other hand, plugging in the average to fill the gap for Zambia might seem reasonable given that its near neighbour, Malawi, has a value of 43.7.

The SDGI is calculated by taking the average (arithmetic) of the values of all 17 SDGs, once any gaps have been filled. The result is a value of the SDGI ranging from 0 to 100, and its creators suggest it could be thought of as a percentage degree of attainment:

> The difference between 100 and countries' scores is therefore the distance in percentage that needs to be completed to achieving the SDGs and goals. Sweden's overall Index score (85) suggest that the country is on average 85% of the way to the best possible outcome across the 17 SDGs.
>
> *(Lafortune et al., 2018, p. 8)*

Thus, the SDGI has a straightforward and intuitive feel to it; a measure of progress towards the ultimate goal of sustainable development.

TABLE 10.3 Calculation of entries to use for the missing values for SDG 10 ('Reduced inequality') and SDG 14 ('Life below water') for African countries

	SDG 10	SDG 14
Angola		43.1
Benin	36.3	41.5
Botswana	0	
Burkina Faso	78.2	
Burundi	67.2	
Cabo Verde	36.7	46
Cameroon	43.3	47.3
Central African Republic	19.4	
Chad	53.4	
Congo	29.5	52.7
Côte d'Ivoire	46.8	36.3
Democratic Republic of the Congo	59.1	10.7
Djibouti		31.2
Ethiopia		
Gabon	47.1	53.7
Gambia	44.4	48.4
Ghana	58.8	56.8
Guinea	82.6	53.6
Kenya	36.4	49.6
Lesotho	0.4	
Liberia	84	54
Madagascar	31.8	51.9
Malawi	43.7	
Mali	74.1	
Mauritania	86.3	57.2
Mauritius	39.3	52.7
Mozambique	49.3	65.2
Namibia	0	61.4
Niger	81.9	
Nigeria	9.2	39.4
Rwanda	27.4	
Senegal	52.6	41.7
Sierra Leone	69.5	48.1
South Africa	0	59.2
Sudan	65.7	56
Swaziland	0	
Tanzania	60.3	56.5
Togo	39.7	38.1
Uganda	62.1	
Zambia		
Zimbabwe	56.1	
Average	**45.2**	**48.2**

Note: The numbers in the tables are the calculated values for the SDG 10 and SDG 14 components. See the various data gaps in the table. The averages of these two columns provide the values that the creators of the SDGI used to plug the gaps in the table.

Source: Own creation based on information available at https://www.un.org/sustainable development/.

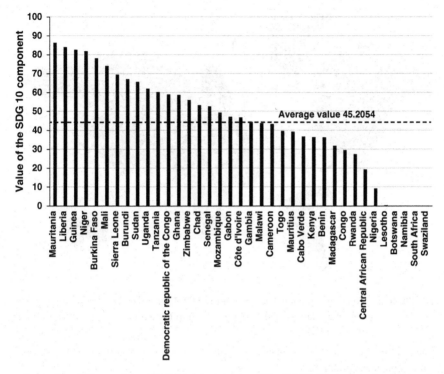

FIGURE 10.1 Values of the SDG 10 component of the SDGI. The values in the graph were used to calculate an average value for all these African countries, which in turn was used to fill in the data gaps (missing values) in SDG 10

Source: Own creation based on data taken from the SDG website (https://www.un.org/sustainabledevelopment/).

The world seen through the lens of the SDGI for 2018 is shown as Figure 10.2. In general, the more developed and wealthier countries of the globe seem to be doing better with regard to SDG attainment than those that are less developed. But this then raises a question as to how the SDGI is related to national incomes? We can assess the latter using GDP/capita (after adjusting for purchasing power), and the result using the GDP/capita for 2016 and the SDGI 2018 is shown as Figure 10.3. The two do indeed appear to be related and the correlation coefficient of 0.84 is high and statistically significant at the 5% cut-off point. It seems that wealth is a reasonable predictor of success when it comes to attainment of the SDGs, something that has logic, as presumably wealthier countries can afford to make the required changes. However, the story is a little more nuanced than that. In Table 10.4 I have presented the correlation coefficients for the 17 components of the SDGI, but I have made sure that missing values have been left as blanks. Twelve of the SDGs do have a positive and significant correlation with income, but for two others ('Responsible consumption and production' and 'Climate action') the relationship is significant

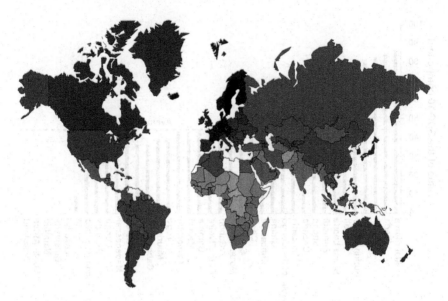

FIGURE 10.2 The world through the lens of the SDGI for 2018. Darker shading equates to higher values for the SDGI, which in turn equates to better attainment of the SDG targets

Source: Own creation based on data taken from the SDG website (https://www.un.org/sustainabledevelopment/).

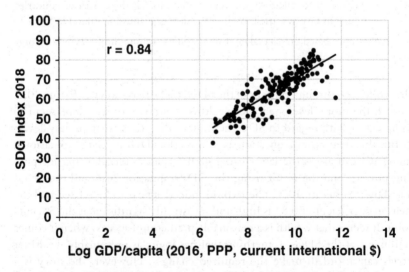

FIGURE 10.3 SDGI (2018) related to logarithm of GDP/capita (adjusted for purchasing power)

Source: Own creation based on SDGI data taken from the SDG website (https://www.un.org/sustainabledevelopment/) and GDP data available from the World Development Indicators website (http://datatopics.worldbank.org/world-development-indicators/).

TABLE 10.4 Correlation between SDGI components and logarithm (base e) of GDP/ capita (2016, PPP, international dollar)

SDG number	Title of SDG	Correlation coefficient (r)
1	No poverty	0.71
2	Zero hunger	0.76
3	Good health and well-being	0.89
4	Quality education	0.79
5	Gender equality	0.58
6	Clean water and sanitation	0.42
7	Affordable and clean energy	0.83
8	Decent work and economic growth	0.73
9	Industry, innovation and infrastructure	0.86
10	Reduced inequality	0.26
11	Sustainable cities and communities	0.66
12	**Responsible consumption and production**	**−0.78**
13	**Climate action**	**−0.31**
14	**Life below water**	**0.12**
15	**Life on land**	**−0.09**
16	Peace and justice strong institutions	0.70
17	**Partnerships to achieve the goal**	**−0.09**

Note: Lighter shaded cells: statistically significant, but negative, values of the correlation coefficient. Darker shaded cells: Values of the correlation coefficient that are not statistically significant at the traditional 5% cut-off point.

but negative. For the latter two SDGs, higher income results in poorer performance – and there is also a logic here, as it is the wealthier countries that will have less responsible consumption and production and will also contribute the most to greenhouse gas emissions. For three other SDGs there would seem to be no apparent correlation with income.

Given that the SDGs represent the major framework in play up until 2030 for the global attainment of sustainable development, and given the breadth of issues and indicators in the SDGs, it seems reasonable to assume that the SDGI will achieve a high degree of prominence. For all its faults, and in some respects they are no different than we find with any index, the SDGI may well become the one index that rules all others and be the final word in the ever-crowded book on indices. Time will tell whether this will be the case, but the SDGI certainly embodies a direction of travel that will be with us for some decades.

Picture this

The need for good quality data to populate indices has been noted a number of times throughout the book and I do not apologise for raising it again here. The old adage 'rubbish in – rubbish out' applies to indicators and indices as much as it does to any evidence that is meant to inform decisions and action. But there is another aspect to quality with regard to indices that receives very little attention from index-geeks such as myself, but which is arguably the most important of all. The numerical indices covered in this book, as well, I would guess, as the indicators associated with the targets set within the MDGs and SDGs, are probably not the stuff we imagine to be part of our everyday lives. They cover important aspects of humanity, that is for sure, and it is hard to imagine anyone not caring about poverty and happiness although I suppose the picture will be more mixed when it comes to the state of the environment or corruption. After all, those who benefit from corruption may have very mixed views, to put it mildly, about whether it is a good or bad thing! Perspectives on corruption may depend upon whether you are paying the cost or receiving the benefit, and the pernicious and longer-term damage from corruption for society can easily be ignored for short-term personal gain. But in this book, I have assessed all these in terms of numerical and quite technical-looking indicators and indices, even though I have purposely not delved all that deeply into the mechanics of most of them. But while we may not recognise the Human Development Index as something we encounter in our everyday lives, we do, nonetheless, use indicators all the time – but we do so using clues we gain via our senses. We may not keep values of the headcount ratio or Gini Index in our heads, but we all know poverty and inequality when we see it. Similarly, we may not keep tables of the Environmental Performance Index readily to hand, but we know what a degraded environment looks like. This more qualitative perspective on indicators was discussed at some length in Chapter 2 and I return to it here partly to reinforce the point that indicators do play an important part in our lives, and partly to help set the ground for the final index I will cover in this book: The Night Light Development Index (NLDI).[2]

The pictures images in Figures 10.4 and 10.5 illustrate the point about indicators as visual clues. In the photograph in Figure 10.4 we see the signs that all is not well with the environment, including plastic waste, pollutants being released into the atmosphere and extensive deforestation. The pictures appear to be unambiguous, and we do not need an ESI or EPI to tell us that there are environmental problems in those scenes; we can see them for ourselves. In Figure 10.5 we see two photographs of shanty towns in very different parts of the world that have many indications we would normally associate with poverty. In the left-hand image, taken in Kathmandu, the capital city of Nepal, we can see the densely packed, yet flimsy nature of the buildings, put together with a variety of materials that are unlikely to provide much protection from severe weather, and the indications are that these buildings may not have adequate services such as sanitation. In the right-hand image we can also see the signs of poverty, but this picture was taken in one of the wealthiest cities in China – Xiamen. In both pictures we can see the nearby presence of water that looks polluted, and there appear to be real dangers stemming from water-borne diseases and storm damage to the buildings. This brings out one of the facets often associated with poverty – vulnerability – a point we covered in Chapter 6. We do not need a numerical index of vulnerability to tell us that the

FIGURE 10.4 Three images of environmental degradation. Top left: Plastic pollution in water (Kolka, Latvia). Bottom-left: Deforestation. Right-hand image: pollutant release into the air from industry

Source: Pixabay (pixabay.com).

FIGURE 10.5 Two images of poverty. Right-hand image: Xiamen, a town in China with a strong economy. Indeed, the city is considered to be one of the wealthiest in China. But note the presence of skyscrapers in the background of the image and the existence of poverty alongside great wealth

Source: (left-hand image) Max Pixel (www.maxpixel.net). Slums in Kathmandu, Nepal.

Source: (right-hand image) Wikimedia Commons, Peter van der Sluijs, Poverty and slums in Kathmandu in Nepal (GNU Free Documentation License, Version 1.2).

people who live in these places may be vulnerable to all sorts of stresses and shocks, and neither do we need indices of poverty to tell us that many of the households may be living below the poverty line; whether national or international. Maybe the pictures also tell us something about the lack of opportunity that the people living in those shanty towns might face, even if both photographs were taken in relatively wealthy cities. Perhaps this is due to a lack of employment opportunity that pays a good wage. And we do not need to be told that the HDI for that scene is a certain number in order for us to be able to appreciate the plight of the people who live there. In one of the pictures we can also see a marked juxtaposition of wealth, as represented by the apartment blocks in the background, and the poverty in the foreground. We do not need a Gini Index to tell us that the scene screams inequality; we can see it with our own eyes. Thus in just these few pictures we can already see many of the facets covered by the quantitative indicators discussed in this book. More important, perhaps, is that the way we have used the information in those pictures is the way in which most of us use 'indicators'; not as numbers but as sensory clues and signals that tell us about how others survive and live. These 'qualitative' indicators are the ones we use all the time. We make everyday decisions based on them, albeit in conjunction with other sensory information that we digest without even thinking about it. Indices should not dehumanise the world.

There are well-established methods for analysing pictures and videos and these come under the broad heading of 'content analysis' (CA) (see Rose, 2011 for an introduction in using CA for visual material). It would be possible, for example, to identify a set of indicators to look for in visual material and then analyse them to see whether those indicators are present. This *a priori* approach to the analysis of indicators in pictures would be straightforward although it may be limited to a binary decision over absence/presence, rather than having a range of potential values. It may be possible to do something more sophisticated, perhaps by employing a scale of 1 to 5 for the degree of presence of an indicator (e.g. number of dwellings with wooden roofs, or the scale of tree felling in a picture) but this may be stretching matters a little.

However, if qualitative indicators are so widely used and well-established methods, such as content analysis, exist allowing us to analyse such material then why have they attracted so little attention within the indicator community? First, while such qualitative indicators are the mainstay of individual decision-making how do they scale-up to help inform decisions made by policymakers and others? Qualitative indicators can be highly localised. All the pictures in Figures 10.4 and 10.5 were taken at particular place and time, and the question can be asked as to what extent are they representative of a whole country? One could argue, of course, that this very issue also plagues the more quantitative indicators discussed throughout this book. Having a single value for an HDI or HPI for a country might make analysis much easier but is it realistic to expect that a single index can apply to a whole country? Inevitably it must be, at best, an average but even that may be highly debatable. We may just have to accept it as an approximation or a sort of 'representation' of the country and nothing more, but clearly there may be much intra-country diversity that is being ignored. Second, there is the potential for personal bias. What we see as extensive tree felling and environmental degradation in Figure 10.4 may be seen by someone else as but the first stage towards fulfilling an excellent opportunity for develop-ment and hence, ironically, a positive! Similarly, what we may see as indicators of hardship, poverty and lack of opportunity in Figure 10.5 may not necessarily be seen in the same way by those residing in the places shown in the pictures. Some of them may even see such sweeping interpretations as condescending and patronising. It can be easy to assume that the interpretation we place on what we see is the same interpretation shared by everyone, but there are real dangers here. Thus, one of the problems of translating sensory clues into indicators is that one looks through a lens of one's culture, biases, history, views, values, etc. Again, of course, we could argue that this same issue of cultural and other bias is also present in the numerical indices.

But maybe there are other ways in which we can think of translating images into indices and in the next section I provide an example based on imagery derived via satellites.

Pictures to indicators

One example of the translation of images into indicators is provided, perhaps surprisingly, by the Defense Meteorological Satellite Program (DMSP) of the United States. Starting in 1962, the DMSP involved the launch of a series of military satellites, primarily designed to assess meteorological conditions such as cloud cover, surface temperature, rainfall, thunderstorms, hurricanes and typhoons. The idea was to provide up-to-date and extensive assessments of these conditions for the US military across the globe. Aircraft, ships and land-based units clearly require such information and, for the military, it can be critical both in peacetime and during conflict. However, although the DMSP began as a classified programme, it became declassified in 1972 and data became widely available to the scientific community. As of 1998 the DMSP group of satellites were transferred from the military to the National Oceanic and Atmospheric Administration (NOAA). The satellites were designed to collect meteorological data, however one of the features of some of the DMSP satellites was an ability to photograph clouds at night using moonlight, and this ability to capture low levels of light also allowed them to photograph artificial light on the surface of the planet during the night. The result is the image shown in Figure 10.6 of night-time lights (based on night light from 2006). Most of the night light in Figure 10.6 is from towns and cities (street lights, buildings, cars, etc.), though some will come from rural areas. There are other parts of the globe, the poles and deserts, which are largely free of night light. Thus, the night-light picture is largely showing the presence of human habitation and activity.

However, while Figure 10.6 is a striking picture of global night light, the actual data collected by the DMSP satellites is numerical, namely the light

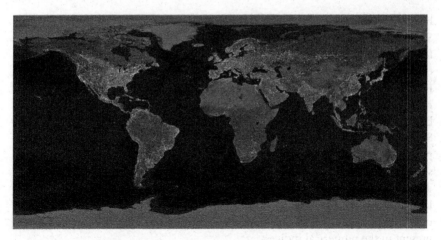

FIGURE 10.6   Global distribution of night light. The photo was derived from satellite imagery supplied via the DMSP of the US

Source: Wikimedia Commons, NASA Earth observatory images by Joshua Stevens, using Suomi NPP VIIRS data from Miguel Román, NASA's Goddard Space Flight Center.

intensity of each pixel equivalent to 1 square kilometre. The light intensity for each pixel was measured in terms of 64 bits; a computer notation representing a scale from 0 (no light) to 63 (brightest light); 0 is included as a value, hence the total of 64 bits. Thus, for a country it is possible to measure the spatial distribution of light within its borders, and it seems reasonable to assume that the extent and intensity of artificial light within a country will reflect the degree of electrification and vehicle ownership, which in turn will be related to 'development'. Hence, wealthier countries may be assumed to have greater generating capacity, a more regular and reliable supply, better distribution networks, including supplies to rural areas, than less wealthy countries and this will result in the presence of more light being seen coming from the land area (urban and rural) of more wealthy countries. In less wealthy countries the electrification, and hence light emission, may be more restricted to cities, even if the supply is irregular, and rural areas may thus appear relatively dark. A similar argument can be made for the degree of vehicle ownership; another potential source of night light. While these assumptions might appear to be somewhat sweeping, there is nonetheless much published evidence to suggest that in less wealthy countries the rural areas are often neglected when it comes to electrification, and governments tend to prioritise scarce electricity supplies to urban and industrial areas. This may be so even if a higher proportion of the country's population live in rural areas relative to urban areas (see, for example, Legros et al., 2009). Rural electrification has often been regarded as an important requirement for development, although perhaps surprisingly, at least for some, it did not make its way into the MDGs other than indirectly via a target to reduce carbon dioxide emissions. But the supply of affordable and clean energy for all does feature strongly in the successor to the MDGS – the SDGs. Indeed, it is the focus of SDG 7:

> At the current time, there are approximately 3 billion people who lack access to clean-cooking solutions and are exposed to dangerous levels of air pollution. Additionally, slightly less than 1 billion people are functioning without electricity and 50% of them are found in Sub-Saharan Africa alone.
>
> *(www.un.org/sustainabledevelopment/energy/)*

Readers interested in the relationship between power supply and development are referred to Barnes (2014) and Rogers and Williams (2015).

Night Light Development Index

Given the assumed linkage between night light, electrification and development set out in the previous section, how can the picture of night lights shown in Figure 10.6 be translated into a development index? This has been attempted and the result is the Night Light Development Index (NLDI), calculated for each country. The details are set out in journal papers written by Ghosh et al. (2010, 2013). While the basis for the translation from the picture of Figure 10.6 to an index is a somewhat tortuous one, it is worth summarising as it does provide points of interest that make the NLDI quite unique when compared to the other indices in this book. The starting point is the sort of spatial grid and distribution of population and night light seen in Figure 10.7, although these data have been made up and are not real; they have been taken from an example provided in Ghosh et al. (2010). In this case there is a 10 × 10 grid comprising 100 squares in total, and the two graphs show the number of people in each square and the

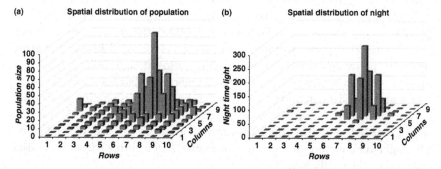

FIGURE 10.7   Spatial grids of population size and night-time light. Each spatial grid comprises 10 rows and 10 columns = 100 squares

Source: Own creation using data from Ghosh et al. (2010, 2013).

'amount' of light from each square. Only a few numbers have been used here for population and light:

Population/square: 0, 1, 2, 3, 4, 5, 10, 15, 30, 50 and 100
Quantity of light/square: 0, 1, 2, 5, 10, 50, 150 and 255

This use of just a few values is, of course, highly simplified but it serves to illustrate the idea. Even from a glance at Figure 10.7 it appears that the two are spatially related, such that a larger population generates more light, but notice how the distribution across space is patchy. The population and night-light distribution grids can be mapped onto each other so we can determine how the two are related. The process is outlined in Table 10.5 and starts with knowing the distribution of population and light across the 10 × 10 grid (columns towards the left-hand side of the table). Thus we can work out that for the 73 squares having a 0 quantity of light each there is a total of 215 people living in those squares. For the 1 square having 1 unit of light there is a total of 10 people and so on up to the 1 square that has 255 units of light and 100 people living there. From this we can work out the proportions of the total population (820) and total quantity of light (1,137 units) corresponding to each of the night-light categories; and from that we can then work out the cumulative values for population and light across the night-light categories. From here it is but a small step to work out the Gini coefficient (as discussed in Chapter 6) for the night-light distribution across population and the result is shown in Figure 10.8. The Gini coefficient is the NLDI and in this example it works out to be 0.46. The larger the NLDI then the greater the inequality in distribution of night light among the population, perhaps reflecting that much of the population lives in places where little night light is produced (e.g. rural areas), while only a relatively small proportion of the population lives in night-light-rich places, which are probably more urbanised areas. If we accept the arguments made in the previous section that better development tends to be equated with better access to a reliable supply of electricity among the population, then the smaller the NLDI the greater the level of development.

While the example above has used made-up data to illustrate the logic of going from population and night light to the NLDI, the process can be applied to real data. The spatial distribution of people is often well-known for many countries, partly as a result of census data but also through estimates, and at a higher resolution we can also get matching night-light data for that same space. The night-light data are in the forms of pixels each corresponding to a square kilometre (1,000 m × 1,000 m) with each pixel having a value (0 to 63) that corresponds to light intensity; the higher the value the greater the light intensity. The night-time picture of the planet in Figure 10.6 is formed from these data collected by the satellites in 2006.

The process for finding the NLDI as shown above may appear to be unnecessarily convoluted. Surely a simpler and more resonant approach would be to add up all the night light produced by a country and divide by its surface area

TABLE 10.5 Steps involved in the calculation of the NLDI

People/square	Number of squares	Number of people (people/square × number of squares)	Light/square	Number of squares	Quantity of light (light/square × number of squares)	Number of people for each category of light/square	Proportion of light in each category	Proportion of population in each category	Cumulative light across categories	Cumulative population across categories
0	13	0	0	73	0	215	0	0.262	*0*	*0*
1	13	13	1	1	1	10	0.001	0.012	*0.001*	*0.274*
2	9	18	2	8	16	80	0.014	0.098	*0.015*	*0.372*
3	7	21	5	5	25	55	0.022	0.067	*0.037*	*0.439*
4	7	28	10	4	40	40	0.035	0.049	*0.072*	*0.488*
5	21	105	50	4	200	120	0.176	0.146	*0.248*	*0.634*
10	20	200	150	4	600	200	0.528	0.244	*0.776*	*0.878*
15	1	15	255	1	255	100	0.224	0.122	*1*	*1*
30	4	120								
50	4	200								
100	1	100								
Totals	**100**	**820**		**100**	**1137**	**820**	**1**	**1**		

Note: The data for this illustration are theoretical and based on the distribution of night light and population across a 10 × 10 spatial grid (100 squares in total) as shown in Figure 10.7. The figures in the right-hand columns (cumulative night-time light and population) are shown plotted in Figure 10.8, and it is from here that the NLDI is calculated.

Source: Adapted from Ghosh et al. (2010, 2013).

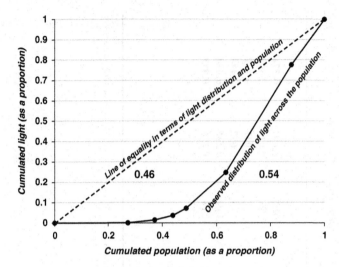

FIGURE 10.8   Observed distribution of night light across the population. The dashed line represents equality in distribution, hence every person produces the same night light. The other line is the observed distribution based on the spatial grids shown in Figure 10.7 and the calculations in Table 10.5. The figures (0.46 and 0.54) are the areas (as a proportion) to the right of the dashed line. In this case the NLDI is 0.46 (equivalent to the Gini coefficient)

Source: Own creation using data from Ghosh et al. (2010, 2013).

or perhaps by population? That would give us a light intensity/area and light intensity/capita, and both would arguably provide a more intuitive index than using a measure of distribution of light amongst a population. Thus, why have the creators of the NLDI opted to use the Gini coefficient as the basis for the NLDI? To help answer this question, I will use two examples of the NLDI calculation for Australia and Nigeria with their night light seen in the highlighted boxes in Figure 10.9. While Australia is a developed country (HDI in 2017 = 0.939), the night-light map of the island continent does suggest a relative absence of light compared with Europe or North America. Nigeria, a country typically classified as 'less developed' (HDI 2017 = 0.532) also seems to have a relatively low level of night light. But the reasons for the relative paucity of night light are different for the two countries. In Australia, Figure 10.9 suggests that the night light is concentrated around the coast, which is where most of the population lives. The inland region of the country, the largest surface area, is arid and the population density is relatively low as a result. For Nigeria, a smaller country by surface area than Australia, the lack of night-time light for much of the country is largely the result of a lack of a consistent and reliable electricity supply. Indeed, for much of the population in rural areas there is no supply of electricity and people have to use other sources such as wood, kerosene or, for those with a generator, diesel or petrol. Hence, the areas that

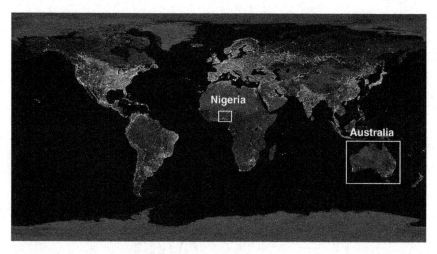

FIGURE 10.9   Night–light picture of the world with Nigeria and Australia highlighted.
Note the concentration of night light around the coastal areas of Australia
while the inland areas of the continent are dark

Source: Adapted from NASA Earth Observatory image by Joshua Stevens, using Suomi
NPP VIIRS data from Miguel Román, NASA's Goddard Space Flight Center.

appear dark are not so because of a lack of people, as is mostly the case in the
inland areas of Australia, but a lack of electricity, and this can be assumed to
relate to less development. If the NLDI was simply based on the quantity of
light coming from a country, perhaps divided by land area or population, then
we may get some widely different results for these two countries. Australia has
more than eight times the land area of Nigeria (7,692,024 km² and 923,768 km²
respectively) while Nigeria has nearly eight times the population of Australia
(25 million and 191 million respectively). If land area was used as the denominator then Nigeria may well come out with a higher value for night light per
area than Australia, while if population was the denominator then Nigeria
may well emerge with a much lower night light per capita than Australia.
Therefore, while population would seem to be a more sensible comparison
to use than land area, there does seem to be a much stronger match between
where the night light is coming from and where the majority of the people live
in Australia relative to Nigeria. It is this logic that drove the decision by the
creators of the NLDI to use distribution of night light amongst the population rather than simple ratios between night light for a country and its area or
population. The calculations of the NLDI for Australia and Nigeria are shown
in Figures 10.10 and 10.11 respectively. The NLDI for Australia is 0.64 while
for Nigeria it is 0.91; indicating that Australia has a more equitable distribution (lower Gini coefficient) of night light amongst its population than does
Nigeria. If we follow the logic set out here and in the previous section, then we
could say that Australia is more developed than Nigeria.

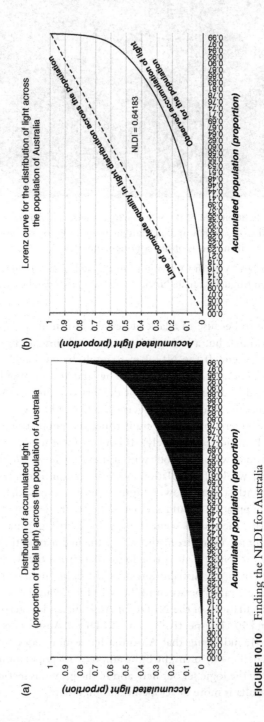

FIGURE 10.10 Finding the NLDI for Australia

Source: Own creation based on data available from https://www.ngdc.noaa.gov/eog/dmsp/download_nldi.html.

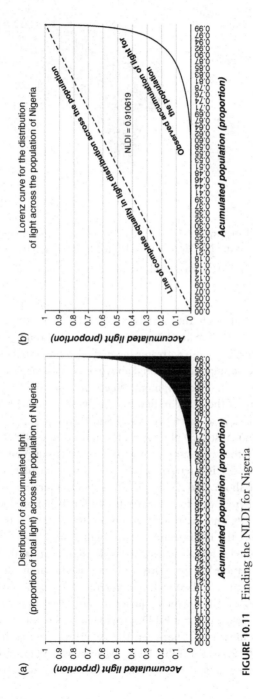

FIGURE 10.11 Finding the NLDI for Nigeria

Source: Own creation based on data available from https://www.ngdc.noaa.gov/eog/dmsp/download_nldi.html.

FIGURE 10.12 Global map of the NLDI for 2006. Darker shading equates to higher values of the NLDI, which is claimed to imply less development

Source: Own creation based on data available from https://www.ngdc.noaa.gov/eog/dmsp/download_nldi.html.

The world seen through the lens of the NLDI (for 2006) is shown in Figure 10.12. Darker shading in this picture equates to countries with higher values for the NLDI; those with a greater inequality in the distribution of night-time light across the population. As noted above, it is assumed that higher values of the NLDI equate to less development. The story presented in Figure 10.12 is in line with many of the other indices in this book. The NLDI is especially high in sub-Saharan Africa, part of Asia and South America. It is low in North America, Europe and Australia/New Zealand. Indeed, the NLDI can be compared to a host of other indicators using the methods outlined in Chapter 9 and the interested reader is referred to two papers in particular, which present a number of these (Ghosh et al., 2010, 2013). From the picture of the world in Figure 10.12, the obvious one to compare the NLDI (based on 2006 data) with is the HDI from 2006, and the results of this are shown in Figure 10.13a. The pattern is a clear one; the HDI declines with increasing NLDI and the correlation coefficient is −0.84, a figure that is statistically significant at the usual cut-off point of 5%. The relationship is negative, but we need to remember that lower levels of the NLDI equate to better development. Thus, as the NDLI decreases then the HDI increases; all of which equates to more development. Hence, the relationship between the HDI and NLDI provides a very consistent pattern. The other interesting comparison is between the NLDI and the Gini Index based on income for 2006 (Figure 10.13b). Here the story is perhaps more surprising as the two indices do not appear to be related that well. In fact, the correlation

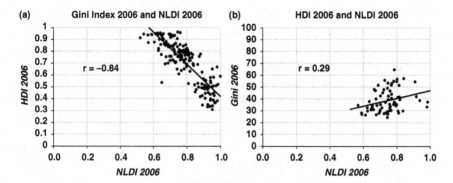

FIGURE 10.13 Correlation of the NLDI with the HDI and the Gini Index of inequality. Each dot in the scatterplots represents a single country

Source: Own creation using NLDI data available from https://www.ngdc.noaa.gov/eog/dmsp/download_nldi.html.

coefficient of 0.29 is also statistically significant (albeit only just) at the 5% cut-off point, but the relationship is arguably not as strong as one would expect to see. Why is this apparently the case? One issue, of course, is that the poor, in terms of income in a population, may not necessarily reside in places that are poor in night light. In fact, the poorest segments of a population may often live in urban areas that produce a lot of light, and it is not inconceivable that a country may have a high inequality in terms of income yet at the same time have a low inequality in terms of night-light distribution amongst the population. Thus, as odd as it may at first seem, it is not unreasonable to expect a weak match between inequality in distribution of night light and inequality of income.

Throwing light on environmental sustainability

How does the NLDI relate to sustainability? The authors of a number of papers on the NLDI relate it to the ecological footprint (Chapter 4) and the pattern is much the same as for the HDI; countries with a higher EF tend to have lower values of the NLDI (Ghosh et al., 2010, 2013). As with the HDI, this result is probably not that surprising as wealthier countries (higher GDP/capita) also tend to have better electricity generation, distribution, vehicle ownership, etc. that all contribute to night light, and the EF is related to GDP/capita, as we have seen in Chapter 9.

However, there are two other environmental indices that have been mentioned in this book: The Environmental Performance Index (EPI) and its predecessor the Environmental Sustainability Index (ESI), both of which were covered in Chapter 5. How does the NLDI relate to them? This is an aspect not covered by the team who put together the NLDI, but it is still an interesting question to ask, especially as the production of electricity still depends heavily on fossil fuel and thus is linked to the release of greenhouse gasses such as carbon dioxide. A plot of NLDI against the EPI 2006 is shown in Figure 10.14a (correlation coefficient = −0.74) and the story is similar to that of the HDI versus NLDI relationship discussed above. As the NLDI increases then the EPI declines, suggesting that environmental performance declines with greater inequality in the distribution of night light amongst the population. Given that we have already seen how the EPI and HDI are both related to income per capita (Chapter 9), then it is perhaps not unsurprising that a significant relationship between NLDI and HDI should suggest a significant relationship between NLDI and EPI.

Perhaps more surprising than the significant link between NLDI and EPI, is that the ESI 2005 does not appear to have a strong relationship with the NLDI, at least not as strong as we see for the EPI (Figure 10.14b). Care does need to be taken here as the correlation coefficient of −0.25 for the NLDI and ESI is statistically significant at the cut-off point of 5%, and the negative nature of the relationship again suggests that the ESI increases with more development (i.e. lower NLDI). We have already noted in Chapter 9 that the EPI and ESI both have a positive relationship with national wealth so perhaps it is not surprising to find that they also have the relationship they do with the NLDI. But why is it stronger for the EPI than the ESI? We may be able to glean some answers for this based upon the five components of the ESI 2005 and how they relate to the NLDI.

Details of the ESI were covered in Chapter 5, and in that chapter the five components of the index were set out as follows:

- Environmental systems (SYSTEM)
- Environmental stresses (STRESS)
- Human vulnerability (VULNER)
- Social and institutional capacity (CAP)
- Global stewardship (GLOBAL)

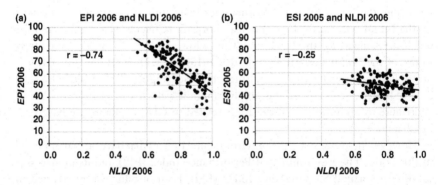

FIGURE 10.14   Correlation of the NLDI with the EPI and the ESI. Each dot in the scatterplots represents a single country

Source: Own creation using NLDI data available from https://www.ngdc.noaa.gov/eog/dmsp/download_nldi.html.

The first two in the list, environmental systems and environmental stresses, respectively, cover the physical quality or state of the environment (SYSTEM) and the release of pollutants, etc. into the environment that may degrade its quality (STRESS). Human vulnerability covers aspects of human health and welfare that are negatively impacted upon by the environment. The last two in the list, social and institutional capacity and global stewardship, are very broad and cover a wide range of issues that broadly come under the institutional, investment and policy domains. The ESI 2006 is, in essence, an amalgamation of these five themes, and with each component, as with the ESI, the logic used by the creators of the index is that the higher the value then the better the contribution towards environmental sustainability. Thus, higher values of SYSTEM and STRESS equate to better environmental quality and less pressure on the environment in terms of pollutant release, etc. Maps of the world through these five components of the ESI 205 are presented in Figure 5.9 of Chapter 5. However, when we look at the relationship between the five components of the ESI 2005 and the NLDI we begin to see complex patterns emerging (Figure 10.15). For the SYSTEM and GLOBAL components there is no apparent relationship with the NLDI. All we see is an apparently random scatter of points across the spaces in each graph. For the other three components there do seem to be relationships with the NLDI, although the stories are more complicated that the seemingly more straightforward linear relationships of Figures 10.13 and 10.14. The relationships for STRESS, VULNER and CAP are 'curved' and go in a variety of directions. For STRESS, the highest values (e.g. less stress placed on the environment via pollution, etc.) equate to countries with the lowest development (highest NLDI), and this has a logic as these countries will, in general, have lower degrees of industrialisation. We would also expect countries that do badly with STRESS (e.g. the industrialised and generally richer countries) would have higher values for the NLDI. For VULNER and CAP we seem to have declines in their values as the NLDI increases (i.e. level of development declines). This appears to be logical, as VULNER would be expected to be worse in the less developed world, and, as I have noted before, CAP does have a link with national wealth and so would also be expected to be lower with less development. What is more interesting with these three components is not so much that the direction of travel with regard to the NLDI makes sense. but that the relationships are more curved than linear. STRESS levels off with increasing NLDI; it would seem that beyond a certain point less development does not bring further improvements in environmental STRESS. CAP appears to hit a 'floor' in the graph, where further declines in CAP do not happen as NLDI increases further, and this perhaps reflects a reality whereby even the poorest countries will still have a level of investment in their institutions, etc. With VULNER we seem to have a situation where even with the wealthiest countries there is a 'ceiling of human vulnerability' above which countries cannot go (at least implied with the 2005 data). Hence there always seems to be a degree of vulnerability in any society, even the richest.

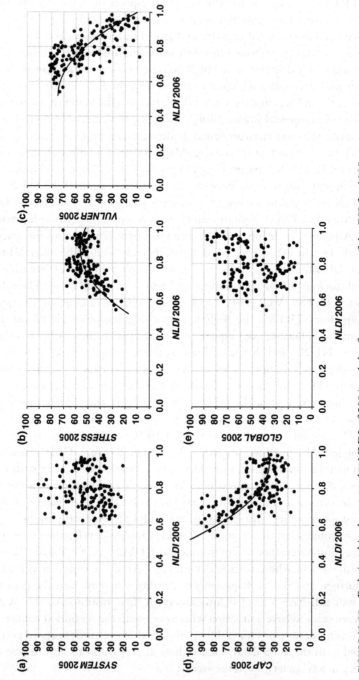

FIGURE 10.15 Relationships between the NLDI of 2006 and the five components of the ESI for 2005

Source: Own creation using NLDI data available from https://www.ngdc.noaa.gov/eog/dmsp/download_nldi.html.

While the apparent linkage between some of the components of the ESI and NLDI may well be mediated by a common relationship with national wealth, perhaps the most intriguing aspect is the potential of using data from satellites to assess aspects of human life that we would not expect to see from orbit. While the notion of using satellites to observe changes in land use is well-established, who would have thought that the CAP and VULNER components of the ESI would also be assessible via satellite imagery? Indeed, this technology offers us a whole new way of thinking about the population of indices (Andries et al., 2018), and indeed there have been efforts to use night light from satellites to assess issues such as human rights (Li et al., 2017), corruption (Hodler and Raschky, 2014), and even the incidence of breast cancer (Rybnikova and Portnov, 2017).

Conclusions: Some thoughts on the future of indicators and indices

Whether we like it or not, indicators and indices will probably be with us for as long as the human race survives. There will always be a need to condense complexity so as to help inform decisions, and while the technology for collecting data may change, as will the indicators, I am confident in my prediction that, as tools, they will always be with us. We may not see the Human Development Index or the Corruption Perception Index in 30 years' time, although these indices have stood the test of time very well until now, but there will be others that will do much the same in a variety of spheres of interest, and we will also have new ways of populating them. We may eliminate poverty and corruption so perhaps we no longer need indicators for them, but who knows what other areas of concern will emerge and they will need indicators and indices to help us manage them.

The NLDI provides an intriguing example of how technology can influence the future of indicators and indices. The correlation of the NLDI with a number of the indices in this book raises the possibility that satellite-based Earth observation may be able to address one of the key issues that often emerges with indicators: The timely availability of good quality data. This issue has been mentioned a great deal throughout this book and is one that still remains challenging, especially for the many developing countries of the world that cannot devote the resources required to routinely collect and verify data that we need to put into indicators. Even if only some of these data can be collected by satellites then it would help. The estimation of night light, which is at the heart of the NLDI, is founded on satellite technology that has its roots in the early 1960s, but this technology has moved on rapidly since the very first satellite (Sputnik) was launched on 4 October 1957, the year I was born. Sputnik was only capable of sending beeps, and nothing more, until it burned up in the Earth's atmosphere in 1958. It could not take photographs, although it did cause quite a stir in military circles as the potential for what had happened became clear. The notion that in the near future a satellite in permanent orbit around the Earth could take high resolution pictures, in colour, of a nation's infrastructure and defences as well as listen to radio signals was not lost on the leaders of the great powers. Sputnik spurred what we now call 'the space race', and while it is the manned missions that have understandably grabbed attention since Yuri Gagarin completed the first orbit of the Earth in 1961, it is the development in satellite technology that is perhaps most relevant for my story. Progress has been phenomenal. I am putting to one side the satellites used for military purposes, of which there are many, including the Defense Meteorological Satellite Program (DMSP) of the United States before the programme was declassified, and instead I am focussing on those that can be used for more peaceful purposes. Today there are satellites with resolutions of less than 0.5^2 per pixel (Andries et al., 2018), a resolution that would allow the desk I am working on when writing this to be seen. Satellites can work with a variety of wavelengths of light and even use a laser beam to measure heights of buildings, trees and landforms. The technology on board satellites can even detect if a plant is infected by pests or disease, or whether it is missing a nutrient, and can also tell us whether a body of water is being polluted. And for more socio-economic indicators, there has been much work to see whether they can also be assessed via satellite-based technology, even if it means using a proxy indicator such as the night-light data discussed above (Andries et al., 2018).

The potential of new technology, such as satellite-based systems and, indeed, other remote sensing systems such as aircraft and drones, is timely given the new game in town – the Sustainable Development Goals (SDGs) and their suite of targets and indicators – that will take us to 2020 and perhaps beyond. The SDGs follow on from the Millennium Development Goals (MDGs) but are intended to be truly global; almost every country has signed up to the SDGs while the MDGs were primarily seen as something for the less developed parts of the world. The SDGs are ambitious but also very necessary for a planet that is in real danger of turning toxic for humankind. At the heart of the SDGs is a large suite of indicators, but one of the key constraints is availability of enough good quality data available on a timely basis. These data are often collected via surveys, but it takes time to collect the data in this way and, as noted in Chapter 7, there are issues about sample size, data quality, etc. to consider; and this can also be a resource-demanding exercise. Earth observation can potentially provide an alternative means of collecting data for indicators and indices. However, while new technology, such as that available via satellites and other machines, provides great potential, it is important not to get carried away and think that they can provide the only answer. That is simply not the case. Earth observation provides us with some new ways for populating indicators and indices, and perhaps provides us with a way of complementing and confirming other data collected on the ground, but it is certainly not a panacea. It also does not address the central point that has been raised so many times in this book. Having indices is one thing, but using them to help bring about a positive change in all our lives as well as the lives of future generations is something else. For the latter we need people who can make that difference; indices will not achieve it by themselves.

However, as the SDGs have been signed up to by almost every country on the planet then they will be a major driving force over the coming decade or so until 2030, and I am optimistic that the suite of indicators used in the SDGs along with the SDGI as a headline index to help highlight progress (or not, as the case may be), will continue to play a major role.

This book has made use of maps of the world to show what the planet would look like if we could 'see' the indicators and indices. The maps bring into sharp relief the geographical differences we see across the globe. The global pictures are different but they also provide a consistency of message. In broad terms, national wealth assessed via GDP tends to bring better human development (HDI), less corruption (CPI), less poverty, greater equality (Gini Index), less vulnerability to climate change, better environmental sustainability (ESI) and performance (EPI), more happiness (Happiness Index and Happy Planet Index) and more sustainable development (SDGI). This is a generalization, of course, and is not true for all countries. There is also the issue of cause–effect to consider – does increased wealth cause less corruption or is it the other way around? But from the many graphs that have been presented, it would appear that national wealth does seem to be at the heart of the index relationships we have set out in the book. Indeed, it is not that hard to imagine why this would be the case. National wealth brings better enforcement of laws to control criminal behaviour and environmental degradation, as well as better incomes (on average), healthcare systems, education, technology and so on, and it is not hard to imagine that this would translate into more happiness. But national wealth also seems to bring with it a higher ecological footprint, a greater impact on the planet's resources. So, is that it then? Does it all really come down to national wealth? Is improving wealth all we need to focus on and everything else will follow? Well clearly not, and the details really do matter. For example, there are countries with similar levels of national wealth but very different ecological footprints. Indeed, the latter can often be hidden or minimalised as we force the data to fit straight lines by using transformations such as logarithms. This is perhaps where indicators and indices come into their own; not so much in the generalities we see in the many scattergraphs and correlation coefficients of Chapters 9 and 10, but the exceptions – the odd ones out – the countries that are not on the neat lines that we can fit to the data. The danger as researchers is that we focus perhaps too much on the trends – the regression lines – as these give us the big picture for academic publications and provide a neat reduction of complexity to a simple message, but nowhere near enough on the data points that are outside of that. Maybe it is on the outliers where we need to place much more attention.

Finally, what the numerous maps in this book do not tell us are the underlying reasons for the geographical disparities that we see. Why is it that after so many years of development intervention we still see the same differences? Why is it that the countries of Africa still dominate the lower reaches of the HDI league table? To be sure there have been many changes since the end of the Second World War, with the rise of China to become the second biggest economy in the world being perhaps the most apparent, and these will continue. But it is as well to remember that using indices to capture rank in country league tables is one thing, and arguably is the easy part; explaining any change so we can learn the lessons is something entirely different and far more challenging. But at least the indices give us a starting point, and we do have to start from somewhere.

Notes

1 A wealth of information on the Sustainable Development Goals is available at the UN website, including datasets of technical details for download: https://www.un.org/sustainabledevelopment/.
2 The night-light datasets at the heart of the NLDI are freely available and can be found here: https://www.ngdc.noaa.gov/eog/dmsp/download_nldi.html.

References

Andrews, N, Khalema, N E and Assié-Lumumba N'D T (eds.) (2015). *Millennium Development Goals (MDGs) in Retrospect: Africa's Development Beyond 2015.* Springer, Dordrecht.

Andries, A, Morse, S, Murphy, R, Lynch, J, Woolliams, E and Fonweban, J (2018). *Translation of Earth Observation Data into Sustainable Development Indicators: An Analytical Framework.* Sustainable Development. https://onlinelibrary.wiley.com/doi/10.1002/sd.1908.

Barnes, D F (2014). *Electric Power for Rural Growth: How Electricity Affects Rural Life in Developing Countries.* 2nd edition. Energy for Development, Washington, DC.

Cimadamore, A, Koehler, G and Pogge, T (2016). *Poverty and the Millennium Development Goals: A Critical Look Forward.* Zed Books, London.

Economist (2015). Leader. The 169 commandments. The proposed sustainable development goals would be worse than useless. https://www.economist.com/leaders/2015/03/26/the-169-commandments.

Ghosh, T, Anderson, SJ, Elvidge, C D and Sutton, P C (2013). Using night time satellite imagery as a proxy measure of human well-being. *Sustainability* 5, 4988–5019.

Ghosh, T, Powell, R, Elvidge, C D, Baugh, K E, Sutton, P C and Anderson, S (2010). Shedding light on the global distribution of economic activity. *The Open Geography Journal* (3), 148–161.

Hodler, R and Raschky, P A (2014). Regional favoritism. *Quarterly Journal of Economics* 129 (2), 995–1033.

Lafortune, G, Fuller, G, Moreno, J, Schmidt-Traub, G and Kroll, C (2018). *SDG Index and Dashboards.* Detailed Methodological paper.

Legros, G, Havet, I, Bruce, N and Bonjour, S (2009). *The Energy Access Situation in Developing Countries. A Review Focusing on the Least Developed Countries and Sub-Saharan Africa.* UNDP and World Health Organization, New York.

Li, X, Li, D, Xu, H and Wu, C (2017). Intercalibration between DMSP/OLS and VIIRS night-time light images to evaluate city light dynamics of Syria's major human settlement during Syrian Civil War. *International Journal of Remote Sensing,* 38 (21), 5934–5951.

Rogers, J and Williams, S P (2015). *Lighting the World: Transforming Our Energy Future by Bringing Electricity to Everyone.* St. Martin's Press, New York.

Rose, G (2011). *Visual Methodologies: An Introduction to Researching with Visual Materials.* 3rd edition. Sage, London.

Rybnikova, N A and Portnov, B A (2017). Outdoor light and breast cancer incidence: A comparative analysis of DMSP and VIIRS-DNB satellite data. *International Journal of Remote Sensing*, 38 (21), 5952–5961.

Further reading

Bell, S and Morse, S (eds.) (2018). *Routledge Handbook of Sustainability Indicators.* Routledge, Abingdon, UK.

Emery, W and Camps, A(2017). *Introduction to Satellite Remote Sensing. Atmosphere, Ocean, Land and Cryosphere Applications.* Elsevier, Amsterdam, The Netherlands.

For a broad introduction to content analysis, although focusing primary on transcripts from interviews, please see the following books all published by Sage:

Krippendorff, K (2013). *Content Analysis: An Introduction to Its Methodology.* 3rd edition. Sage, Los Angeles, CA, and London.

Miles, M B, Huberman, A M and Saldaña, J (2014). *Qualitative Data Analysis: A Methods Sourcebook.* 3rd edition. Sage, Los Angeles, CA, and London.

Neuendorf ,K A (2017). *The Content Analysis Guidebook.* 2nd edition. Sage, Los Angeles, CA, and London.

INDEX

Note: Page numbers in *italics* indicate figures. Page numbers in **bold** indicate tables.

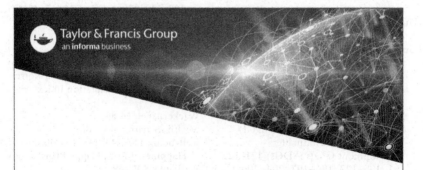